AF739068

Palgrave Studies in Impact Finance

Series Editor
Mario La Torre, Department of Management, Sapienza University of Rome, Rome, Italy

The *Palgrave Studies in Impact Finance* series provides a valuable scientific 'hub' for researchers, professionals and policy makers involved in Impact finance and related topics. It includes studies in the social, political, environmental and ethical impact of finance, exploring all aspects of impact finance and socially responsible investment, including policy issues, financial instruments, markets and clients, standards, regulations and financial management, with a particular focus on impact investments and microfinance.

Titles feature the most recent empirical analysis with a theoretical approach, including up to date and innovative studies that cover issues which impact finance and society globally.

More information about this series at
https://link.springer.com/bookseries/14621

Carlo Bellavite Pellegrini · Laura Pellegrini · Massimo Catizone
Editors

Climate Change Adaptation, Governance and New Issues of Value

Measuring the Impact of ESG Scores on CoE and Firm Performance

palgrave
macmillan

Editors
Carlo Bellavite Pellegrini
Department of Economic Policy
Catholic University of the Sacred Heart
Milan, Italy

Laura Pellegrini
Department of Economics and Finance
University of Bari
Aldo Moro, Italy

Massimo Catizone
Catholic University of the Sacred Heart
Milan, Italy

ISSN 2662-5105 ISSN 2662-5113 (electronic)
Palgrave Studies in Impact Finance
ISBN 978-3-030-90114-1 ISBN 978-3-030-90115-8 (eBook)
https://doi.org/10.1007/978-3-030-90115-8

© The Editor(s) (if applicable) and The Author(s), under exclusive license to Springer Nature Switzerland AG 2022
This work is subject to copyright. All rights are solely and exclusively licensed by the Publisher, whether the whole or part of the material is concerned, specifically the rights of translation, reprinting, reuse of illustrations, recitation, broadcasting, reproduction on microfilms or in any other physical way, and transmission or information storage and retrieval, electronic adaptation, computer software, or by similar or dissimilar methodology now known or hereafter developed.
The use of general descriptive names, registered names, trademarks, service marks, etc. in this publication does not imply, even in the absence of a specific statement, that such names are exempt from the relevant protective laws and regulations and therefore free for general use.
The publisher, the authors and the editors are safe to assume that the advice and information in this book are believed to be true and accurate at the date of publication. Neither the publisher nor the authors or the editors give a warranty, expressed or implied, with respect to the material contained herein or for any errors or omissions that may have been made. The publisher remains neutral with regard to jurisdictional claims in published maps and institutional affiliations.

Cover illustration: Galeanu Mihai

This Palgrave Macmillan imprint is published by the registered company Springer Nature Switzerland AG
The registered company address is: Gewerbestrasse 11, 6330 Cham, Switzerland

Carlo Bellavite Pellegrini: to Giovanni Bazoli

Laura Pellegrini: to my son, Matteo.
For a better World

Massimo Catizone: to Maria Ludovica, Lucrezia and Liliana

Foreword

There is a growing awareness regarding the need that the corporate management should be increasingly focused on the respect of people and things, in addition to the quality of goods and services produced.

A greater awareness is associated to a growing understanding of the concepts behind the acronym "ESG", a synthesis of a whole way of thinking, whose historical evolution should be recalled.

The "S" stands for "Social" and is the oldest among the three elements: we could say that it was born at the outset of the first industrial revolution, towards the end of the eighteenth century, developing further during the second one in the nineteenth century and giving rise to the theories that would eventually transform into ideologies—Liberalism, Socialism, Marxism—that were characterized essentially by the conflict pertaining to the distribution of the value created by the main actors of the industrial process. At the end of the last century and at the beginning of the current one, the harsh tone of the social conflict weakened, thanks to the general improvement of the level of civilization, and other topical issues gained importance, such as the intolerance to inequalities, the citizens' rights, gender and ethnic disparity, neutrality of sexual identity, realities that have gradually enhanced the significance of the "S" component.

The origins of the "G" of "Governance" are more recent, around the middle of the last century, in response to the need to clearly define and make transparent the rules disciplining the activities of the corporate administration, the relationship among the governing bodies, their independence, their gender composition, the relationship between the ownership—the shareholders—and the management structure, as well as the one with other parties, or stakeholders with a vested interest in the company. This broad subject matter represents the response to the need to safeguard the pillar of the economic system in the Western democracies: the trust of citizens and economic operators in the correct functioning of the market economy under the prerequisite of democracy. At the level of the single enterprise, good Governance builds up, through trust, the reputation and strengthens the competitiveness.

After the tumultuous economic development following the end of the second world war, mainly in Europe, Japan and the United States, at the end of the sixties of the 1900s, in the Western world arose the first serious questions about the "limits to growth" and with them the concerns about the environmental degradation originating from the exploitation of the planet's resources, especially the non-renewable ones (1). From then on, the word "sustainability" became usual in the common language to qualify the wise development of the economic activity and the use of the resources—including also the social aspects and the corporate governance ones—so much so that in the nineties the first "Ethical Funds" were born, their aim being to select their own investments according to sustainability criteria. Despite the difficulty in distinguishing sometimes the substance from the rhetoric, the topic issues had been proposed and they would have increasingly influenced behaviours and choices.

The "E" of "Environmental" enters then forcefully the scene in this century riding the wave of fears concerning the planet's capacity to sustain the economic growth as hitherto known, with the aggravating effect of climate change and specifically, inter alia, by global warming.

These problems have shaken public opinion and the "E" "S" G" factors joined in the synthesis identifying sustainable thinking, starting to become business culture too.

If these are the issues of today agenda, it must be said that they substantially belong to the Western culture, even if unevenly, since the West has so far represented the advanced frontier of the economic development; they do not enjoy widespread appreciation and sensitivity elsewhere in the world, even if other countries have nowadays a higher rate of economic growth than the West.

Since the advent of industrialization, the enterprise has always had to confront with the exogenous entities—similar to those described previously—by taking on these new responsibilities and incorporating them into its operability; inevitably, since enterprises are an integral part of society, sharing its level of civilization, not only the one formalized by law, but also the one in progress in the civil society and public opinion. The business has always done it, acting as a factor of advancement and progress or sometimes, on the contrary, as a factor of rear-guard according to the different awareness of the entrepreneurs and/or the shareholders. The present reality is characterized by the spreading of the awareness towards the issues summarized under the acronym "ESG" among large groups of public opinion, also as a result of media which constitute a factor of pressure on the behaviours and choices of the enterprises, increasing their responsibility, even the social one.

"The Social Responsibility of Business Is To Increase Its profits", is the title of a famous article by Milton Friedman—Noble Prize Winner for Economics in 1976—published in the New York Times Magazine on September 13th 1970, over half a century ago. In the following decades, this sentence would have been brandished as a dialectic weapon, both by those defending its literal interpretation and by those who took position against it, regardless of the contents of the article, less simple than its title... But social responsibility and increase in profits are not necessarily at odds with each other: it's known—and Friedman knew it too—that increased profits do not merely depend on increased incomes and lowered costs, but above all profits depend on investments and, therefore, what matters are not only the short terms quarterly results, but the longer-lasting medium-term performances, as demanded by competitiveness, reputation and trust. The most forward-looking companies of long-standing success—even the ones that from 1990 onward had shared and practised the shareholder value theory —have always invested, in addition to their industrial investments, in Human Capital, in compensation policies, fostered pathways of development of personal growth, improved environmental working condition—a requirement strictly connected to productivity—in training and benefits (e.g. housings and/or recreational centres for workers and their families are experience dating to the beginning of the twentieth century).

All these activities were never for free, they always cost, but they are not expensing detrimental to profits, they are indeed investments that generate returns.

The declaration dated August 19th 2019 of the 181 Chief Executive Officers of The Business Roundtable *"…who commit to leading their companies for the benefits of all stakeholders—customers, employees, suppliers, communities and shareholders…"* seems totally in contrast with Friedman's one of forty-nine years before; it is the expression of a different cultural and historical climate, of course, which shows a higher level of civilization and calls for it.

The two assertions have anyway some overlappings: in order to achieve its mission, the enterprise must reward all the production factors it employs: workers with salary, suppliers and service providers with the cost of their performances, debt financing with interests, the State with taxes and, finally, risk capital with profit; this last one is a residual remuneration available only if and when all other factors have been rewarded, but it has to be rewarded too. If the enterprise is not able to reward it, that means that it employs more resources than it is able to generate and if it doesn't succeed in doing that repeatedly, it fails its mission that is to generate wealth, not to destroy it, causing damages not only to its shareholders, but also to all stakeholders and to the community as a whole that is then forced to take charge of it. Therefore, it is in the interest of all parties that the company remunerates all the production factors thus generating long-lasting wealth, because this is its primary responsibility and only fulfilling it, can enterprises undertake all other responsibilities required by the wider group of stakeholders.

It is not by chance that Leopoldo Pirelli, last of the descendants of the founder, who, from 1956 to 1991, led the family company founded in 1872, by enunciating his ten rules of the "good entrepreneur" stated: *"…the first quality that an entrepreneur must always demonstrate: strive and strive again with every possible means to close good balance sheets. If he doesn't succeed once, try again. If he doesn't succeed repeatedly, he should leave. And if he succeeds, he shouldn't consider himself as God Almighty, but simply someone who, given the profession he chooses, has done his duty"*.

Nevertheless, sustainability was always part of the Pirelli corporate identity even in times when the word was not as usual as it is now: social housing, advanced health care for workers, fair dialogue with the trade unions, also in times of tough conflicts are part of the Pirelli history, as well as its culture of value creation, productivity, profitability. For decades Pirelli stands in the first position in the sustainability rankings.

If the sustainability of its balance sheets is the unavoidable and inescapable responsibility of the company, it's not the only one; there are other responsibilities involved in how the company generates wealth; they consist in the respect for people and things, things being all those resources that the company uses and employs in its production processes, such as energy and the environment.

The role of the enterprise—its evaluation and ethical justification—does not only consist in the capital allocation and in the deployment of production factors by remunerating all of them in order to achieve a specific economic objective but in its ability to meet at the same time the needs of the population of the company's stakeholders whose expectations and values, even if not completely codified, constitute the level of civilisation of a specific geopolitical area in a continually evolving historical time.

This capacity to satisfy a wide range of needs of all stakeholders constitutes a form of immaterial, but very tangible remuneration towards this constituency, as well as a business investment for the company, which generates a return in the appreciation of the value of the company itself, thus rewarding the constituency of the shareholders. This generates a wide circularity that requires management awareness and subtler and more complex metrics of measurement and evaluation in order to be fine-tuned to a context in which reputation and narrative play a more determining role.

Finally, the respect that the enterprises due to people and things pretend to be reciprocated by the respect that the community owes to the role that companies and entrepreneurs play in the process of creating wellbeing for all constituencies and in the progress of the civilization level.

Financial markets reacted positively to this new cultural mood, embracing the "ESG" approach and its criteria in their decisions for selecting investments and have confirmed their appreciation of the sustainable corporate values through lower cost of capital, which reduces volatility; furthermore, the financial markets turned out to be able to modulate the responsiveness of the different factors to each "E", "S", "G" component.

So far, this sensitivity on this subject—as we said before—belongs mainly to the western world culture, but it is not the same in all western countries: Europe seems to be ahead of other areas like United States or Japan.

This book deals in depth with the issues of sustainability in its environmental, social and corporate governance declinations and takes a decisive step, moving from a literary and narrative approach to the scientific one of quantitative measurement of the elements, providing accuracy and rigour to the knowledge base and methodological discipline to its reasoning.

The Second Section of the book is dedicated to the research of the existing correlations between the rating of the "E", "S", "G" components and the relevant economic dimensions, such as the cost of capital and the return on the invested capital in different economic sectors, analysing the importance of sectorial differences towards the meaningfulness of the results, as well as the different weightings of "E", "S" and "G" in the various sectors.

We are just at the beginning of a journey that we can expect to be long and sometimes arduous, but it is already promising: the first indications encourage even the possibility that we could perhaps compose harmoniously the thinking of Milton Friedman, with the one pronounced by the 181 CEO's of the Business Roundtable…

Milan, Italy Enrico Parazzini

Reference

Forrester, J. W., Meadows, D. L., Randers, J., Anderson, A. A., Anderson, J. M., Beherens III, W. W., Naill, R. F., Shantzis, S. B. (1972). MIT Report "The limits to growth", Massachusetts Institute of Technology.

Enrico Parazzini was born in Milan and graduated from the Bocconi University. He is Chairman of Bagheera S.p.A., formerly Managing Director and Deputy Vice Chairman of Prelios S.p.A., Chief Executive Officer of Camfin, Chief Financial Officer of Telecom Italia, Group Controller of Pirelli, General Manager of Bull Italia.

Preface

What do we know about sustainability and its with value and firms?

Perhaps before answering this question a more fundamental question should be answered, namely, what do we mean by value and who are the ultimate beneficiaries of such "Value"? We must admit that we have been asking ourselves these questions many times over the years. Finding a fully satisfactory answer proved to be challenging. Surely there is more than one answer. More than one, as the stakeholders' interest that should be considered.

So, within our studies and empirical research as academics and practitioners, together with Carlo and Massimo, we have dedicated plenty of time to these interesting topics, until one day in October 2019, the idea of this book was born.

"Why don't we write a book that tries to shed some light on these issues by, among other things, providing some initial empirical evidence on the relation between sustainability and value creation?"

So, according to our different and complementary profiles of academics and practitioners, we started thinking through these concepts.

The idea was to lay the basics and more than these for the interpretation of concepts that are so close to each one of us and our lives but not so easy to understand from an economic point of view.

Furthermore, our purpose was to suggest some tools which could demonstrate that the impact in terms of monetary value creation of a sustainable business model is relevant for the company itself but far-reaching benefits are enjoyed by the entire Ecosystem participants.

Sustainable strategies have systemic relevance, affect both debt and equity and have major repercussions on credit risk, liquidity and firm's performance and value. A growing body of evidence is suggesting that the traditional ways of measuring value are obsolete and should be reconsidered. The most obvious example is the failure of the current accounting standards to reflect how the value and long-term viability of any business is affected by the way in which such business affects the value of its most important (off-balance sheet) assets: the natural capital.

Sustainable strategies including Corporate Social Responsibility (CSR) and ESG scores have a direct impact on value and valuations. It is however evident that understanding ESG factors enables a thorough assessment of the risks and opportunities a company faces. In turn, this allows for improved assets allocation, stock picking and risk management. Additionally, ESG analysis leads to improved understanding of how future trends could affect a certain industry or the entire economic landscape for that matter.

Hoping to have succeeded in our purpose, this book tries to contribute to this field from these points of view.

Because sustainability creates long-term value for everyone. Because the shareholder vision has long given way to stakeholder value. This is and will always be more the only starting point that really matter. And together is always better if we look in the same direction.

Milan, Italy — Carlo Bellavite Pellegrini
Bari, Italy — Laura Pellegrini
Milan, Italy — Massimo Catizone

Acknowledgements

Carlo, Massimo and Laura would like to thank Rachele Camacci and Claudia Cannas for their help in formatting the first draft of the book, and Tula Weis, Manikandan Murthy, Shukkanthy Siva, Susan Westendorf, from Palgrave Macmillan and Springernature for their editorial support.

We would also like to thank the co-authors on the studies that directly contributed to the content of this book: Sophia Veronica Barbieri, Rachele Camacci, Claudia Cannas, Raul Caruso, Rocco Cifone, Carlo Montalbetti, Marco Seracini.

About This Book

A large growing literature is nowadays investigating to what extent sustainability drives investors' decisions, and corporate strategies. In some cases, the outcome of such investigations is posing fundamental questions and challenging commonly accepted principles (Mayer, 2018).

In order to fully appreciate what is a sustainable, value-creating investment strategy or business a fundamental re-assessment of the way in which corporates and their mission are defined seems to be required. The traditional elements driving evolution and adaptation as pre-requisites for long-term sustainability and profitability have been brought into question. Is corporate evolution determined by competition or consciousness of purpose? How should traditional ownership structures and corporate governance solutions adapt to reflect a new and more sophisticated corporate mission? Which is the impact?

Although corporate finance has historically researched the determinants of stock and bonds returns and modelling future yields, recently the corporate governance has focused its attention on measuring the impact of non-financial information on listed companies' corporate financial performance. According to the efficient market theory, all new information has the potential to impact the market value of shares. Therefore, it could be stated that the more complete and more reliable the information available, the more accurate is the valuation of the future performance of equity. Even if extra-financial information may not necessarily affect the price of a company's share during normal operations, in cases where reputational or monetarily quantifiable litigation risk exists, investment professionals pay strong attention to the respective information.[1]

This field of study has become more relevant over time due to the increasing attention of the investor. Because different stakeholders have different time horizons but also different objectives. Needless to say, the challenge is to verify whether considering sustainability, environmental and social issues also payoffs in terms of performance and added value to the firm. Because performance can be measured in different ways. Whether it is reasonable to say that such strategies of firms do contribute to the establishment of a more sustainable business context as envisioned in Waddock (2017), there are substantial doubts about the role of ESG in shaping both profitability and firm value [see among others Lee et al. (2018)].

In particular, the ESG scores combine different elements, among which **climate change**—as one of the most prominent and challenging environmental issues facing companies—has a particular relevance for financial markets.

It is foreseeable that companies will have to operate under different conditions in the near future. With this in mind, private capital has an important role to play in preventing and mitigating the impact of climate change.

In economic literature, the search for a relation between Environmental, Social and Governance scores (ESG) and corporate financial performance can be traced back to the beginning of the 1970s.

A first strand of literature focused on the effect that ESG scores have on the cost of capital (equity and debt) and therefore on the related risk, highlighting that companies that have good sustainability standards enjoy a significantly lower cost of debt and cost of equity due to a reduction of the relative risk (Bhojraj & Sengupta, 2003; Schauten & van Dijk, 2011; Bauer & Hann, 2010). Researches have also shown that good corporate governance leads to lower cost of equity (Lima & Sanvincente, 2013), environmental risk management practices, disclosure on environmental policies (El Ghoul et al., 2011), good employee relations and product safety (El Ghoul et al., 2011) lower firm's cost of equity.

[1] That is why companies make an increasing effort to provide investors with disclosures on extra-financial aspects which capture additional dimensions of corporate performance that are not accounted for within financial data.

Further studies also aimed to investigate the effects of sustainability on a company's operating performance. Those studies generally show a positive correlation between the environmental, social and governance topics and operational performance (Fulton et al., 2012; Margolis et al., 2007; van Beurden & Gossling, 2008).

Moreover, other studies have investigated whether this information increases the benefits for equity investors. On the governance dimension, the majority of research suggests that superior governance quality leads to better financial performance (Bebchuk et al., 2009; Cremers & Ferrel, 2014; Gompers et al., 2003). On the environmental dimension of sustainability, eco-efficiency and environmentally responsible behaviour are viewed as the most important factors leading to superior stock market performance (Derwall et al., 2005; Karpoff et al.,2005). Finally, on the social dimension, the literature shows a positive relationship between employee satisfaction and stock market performance (Edmans et al., 2017).

But, when considering aggregate ESG scores more recent literature is already providing researchers with complex evidence (Fatemi et al. 2017; Capelle-Blancard and Petit 2017).

Therefore, limited awareness, lack of data jeopardizing the ability of investors to make risk-adjusted assessments of their expected returns, insufficient harmonized international actions, complexities surrounding valuation methodologies, limited regulatory and tax benefits are only some of the reasons slowing the flow of private capital into **adaptation finance**[2] and the pace of the growth and innovation in the sector. According to this point of view, corporate and systemic governance is therefore a vital component of this ecosystem.

The aim of this book is to investigate sustainability, the relevance of ESG scores, and its impact on firm value and growth in some interesting sectors around the World during the last two decades. The main purposes of this work could be summarized in the following points:

1. Identify and examine the main factors adversely affecting the growth of adaptation finance, including, for example:

[2] Adaptation finance is an ecosystem where the public sector and the private sector should complement each other. This is an essential condition for adaptation finance initiatives to maximise their impact.

a. Inadequate governance system. Lack of a harmonized ecosystem that is conducive to mutually beneficial cooperation between the private and the public sector capable of ensuring sustainable growth, social and financial stability;
b. The limited involvement of the private sector reduces the pace of the development of new financing and insurance solutions;
c. Limited data. The number of databases and relevant indices is rapidly increasing [3] (there exist more than 2000 ESG indices globally), however, available data remains statistically inconclusive:
   i. In the absence of specific risk disclosure requirements (both at issuer level and at country level) predicting returns on adaptation investments remains challenging;
   ii. Pending the production of historical performance data on adaptation finance, it is difficult to argue that adaptation investments are less volatile and less risky than traditional investments. This is a key consideration for any investor acting on an arm's length basis. However, traditional investors (pension funds and investment funds) may assess the issue from a different angle relative to bank investors. For example, discussions on green mortgage loans and energy-efficient loans will remain on a theoretical level as long as there is no reliable data indicating that the risk profile of green mortgage loans and energy-efficient loans justifies a preferential regulatory treatment (i.e. lower risk weights).

2. Deeply focuses on self-consciousness and awareness of corporates and their mission.
3. Argue the need for a harmonized and globally recognized score system for assessing compliance and consistency with adaptation finance objectives. In order for such score system to be recognized and accepted on a global level it will have to be arrived at, based on generally accepted fundamental principles.
4. Make use of case studies to demonstrate the suitability of adaptation finance ratings to predict investment returns.

[3] Goldman Sachs (2019), ESG Index Book.

The book proposes a theoretical and empirical approach towards this topic, based on a comparative assessment of different industries around the World and at the market level. Observations relate to the last twenty years.

The first part of the book analyses the main theoretical and institutional features of the topic and it consists of six chapters devoted to a more accurate study of the conceptual background of these issues, starting from self-consciousness and awareness issues and sustainable business models' issue. Again, we propose an in-depth analysis of climate change challenges and initial policymakers' responses and we focus on ESG scores considered as crucial issues for growth and development. The last chapter is devoted to analysing the main existing literature on a topic related to how to measure the impact of ESG scores on firms' performance and CoE (cost of equity).

The empirical analysis will propose novel models trying to explain the relationship between the score of different systems and risks, with a focus on the impact of these issues on firm value and growth. Following the idea of a comparative approach, we propose some empirical studies in a comparative industry-level approach. To gain the aim, among the others, we will make use of a set of corporate variables i.e. size variables, leverage variables, efficiency variables.

At the end of each empirical chapter, we suggest some concluding remarks, with considerations on empirical evidences and policy recommendations and with an agenda for future research topics.

We use Microsoft Word and Excel. Statistical results from STATA are presented in Word/Excel tables/graphs.

References

Bebchuk, L. A., Cohen, A., & Ferrell, A. (2009). What matters in corporate governance? *Review of Financial Studies, 22*(2), 783–827.

Bhojraj, S., & Sengupta, P. (2003). Effect of corporate governance on bond rating and yields: The role of institutional investors and outside directors. *Journal of Business,* 76(3), 455–475.

Bauer, R., & Hann, D. (2010). *Corporate Environmental Management and Credit Risk.* ECCE Working Paper. University Maastricht. The European Centre for Corporate Engagement.

Capelle-Blancard, G., & Petit, A. (2017). The weighting of CSR dimensions: One size does not fit all. *Business & Society, 56*(6), 919–943.

Cremers, K. J. M., & Ferrell, A. (2014). Thirty years of shareholder rights and firm value. *The Journal of Finance, 69*(3), 1167–1196.

Derwall, J., Bauer, R., Guenster, N., & Koedijk, K. C. G. (2005). The eco-efficiency premium puzzle. *Financial Analysts Journal, 61*(2), 51–63.

Edmans, A., Li, L., & Zhang, C. (2017). Employee satisfaction, labor market flexibility, and stock returns around the world (No. w20300). *National Bureau of Economic Research.*

El Ghoul, S., Guedhami, O., Kwok, C. C. Y., & Mishra, D. R. (2011). Does corporate social responsibility affect the cost of capital? *Journal of Banking & Finance, 35*(9), 2388–2406.

Fatemi, A., Galum, M., & Kaiser, S. (2017). ESG performance and firm value: The moderating role of disclosure. *Global Finance Journal.*

Fulton, M., Kahn, B., & Sharples, C. (2012). Sustainable investing: Establishing long-term value and performance. Available at SSRN 2222740.

Gompers, P. A., Ishii, J., & Metrick, A. (2003). Corporate governance and equity prices. *The Quarterly Journal of Economics, 118*(1), 107–156.

Karpoff, J. M., Lott, J. R., & Wehrly, E. W. (2005). The reputational penalties for environmental violations: Empirical evidence. *Journal of Law and Economics, 48*(2), 653–675.

Lee, J., Graves, S. B., & Waddock, S. (2018). Doing good does not preclude doing well: Corporate responsibility and financial performance. *Social Responsibility Journal, 14*(4), 768–781.

Lima, B. F., & Sanvicente, A. Z. (2013). Quality of corporate governance and cost of equity in Brazil. *Journal of Applied Corporate Finance, 25*(1), 72–80.

Margolis, J. D., Elfenbein, H. A., & Walsh, J. P. (2007). Does it pay to be good? A meta-analysis and redirection of research on the relationship between corporate social and financial performance. Ann Arbor 1001(48109-1234).

Mayer, C. (2018). *Prosperity: Better business makes the greater good.* Oxford University Press.

Schauten M., Van, D., & Dick, J. C. (2011). *Corporate Governance and the Cost of Debt of Large European Firms.* ERIM Report Series Research in Management. No. ERS-2010-025-F&A.

Van Beurden, P., & Gössling, T. (2008). The worth of values–a literature review on the relation between corporate social and financial performance. *Journal of Business Ethics, 82*(2), 407–424.

Waddock, S. (2017). Inequality, dignity and the sustainability challenge. *Journal of Management for Global Sustainability, 5*(1), 63–83.

Contents

Notes on Contributors

Sophia Veronica Barbieri earned a Master's Degree in Economics of Markets and Financial Intermediaries in 2020 at the Catholic University of the Sacred Heart, Milan and subsequently a post-Master's degree in Financial Risk Management at the Politecnico di Milano. She's currently involved in wealth risk management for an investment bank.

Rachele Camacci obtained M.Sc. in Economics of the Financial Institutions and Financial Markets cum laude at Catholic University of the Sacred Heart of Milan. She is a Subject Expert in Corporate Finance and Corporate Governance. She currently carries out research on Green Finance and ESG issues at the Center for Applied Economics Studies (CSEA) of the Catholic University of Sacred Heart of Milan.

Claudia Cannas obtained the M.Sc. in Economics of the Financial Institutions and Financial Markets at Catholic University of Sacred Heart—Milan. She is a Subject Expert on Corporate Governance and Corporate Finance issues. She currently carries out research on Sustainable Finance and ESG issues at the Center for Applied Economics Studies (CSEA) of the Catholic University of Sacred Heart—Milan.

Raul Caruso He is Full Professor of Economic Policy at the Catholic University of the Sacred Heart, Milan where he teaches Peace Economics, Economics of Crime and International Economics. He is Director and member of the Scientific Committee of AssoBenefit and Director of

CESPIC (Centro Europeo di Scienza della Pace, Integrazione e Cooperazione). He was Executive director of NEPS (Network of European Peace Scientists—2009–2019). He is Editor-in-Chief of the Journal of Peace Economics Peace Science and Public Policy. He is the author of books and several articles, at the international and national level.

Massimo Catizone is a member of the scientific committee of CSEA (Centre for Applied Economic Studies) with the Catholic University of the Sacred Heart in Milan, and an investment banker with a leading pan-European investment bank. Prior to investment banking Massimo launched a finance advisory boutique based in London supervised by the FCA, acted as CEO for a European asset management company and covered a number of senior positions in primary international financial institutions, banks and law firms. Massimo graduated from University of Bologna, Italy, studied at the Robert Schuman University, Strasbourg, France, holds a Certificate in Investment Management from the Chartered Institute for Securities and Investments and is qualified as an Italian lawyer and as a solicitor of England and Wales. He is a co-author of The Family Office, Giuffre' Francis Lefebvre (2019) and of articles on sustainability and digitalization.

Rocco Cifone obtained M.sc. in Economics of the Financial Institutions and Financial Markets at Catholic University of Sacred Heart of Milan. He is a senior consultant in Corporate Finance and Corporate Governance with a focus on sustainability issues. In recent years he has collaborated with the Chamber of Commerce of Milan, Monza and Lodi on research into the evolution of Italian companies on governance issues.

Maurizio Dallocchio He is Full Professor of Corporate Finance at Bocconi University of Milan and past Dean of SDA Bocconi School of Management. He is author of books and several articles, at international and national level.

Carlo Montalbetti General Manager of Comieco he is an expert in corporate management, with specialization on the management of industrial cycles related to paper and cardboard recycling and environmental sustainability. Author of several publications of these issues he is a Member of: Symbola Foundation Committee, Strategic Committee of the University of Gastronomic Sciences in Pollenzo, ISVI (Università Bocconi), CSEA Scientific Committee (Catholic U. Milan); of Scuole Civiche Milanesi Board of Directors and of Amerigo Association. He was a member of the Sacbo Board of Directors (Società Aeroporto Civile Bergamo Orio al Serio).

Carlo Bellavite Pellegrini (B.Sc. Economics at the Catholic University of the Sacred Heart of Milan, 1991; Ph.D. in Economics at the Bocconi University of Milan, 1996). Since March 2016 he is Full Professor of Corporate Finance, Faculty of Banking, Finance and Insurance Sciences, Catholic University of the Sacred Heart of Milan where he teaches Corporate Finance and Corporate Governance.

He is Director of the Executive Committee of the Research Centre for studies in Applied Economics. He is fellow of research centres (Centre for Econometric Analysis, Cass Business School, London; Member of the Steering Committee, BAFFI CAREFIN-Centre for Applied Research on International Markets, Banking, Finance and Regulation, Department of Economics, L. Bocconi University; Centre for Research into Cooperation and Non-profit Sector) and Scientific Societies (AIDEA; ECGI-European Corporate Governance Institute; ADEIMF). He was visiting scholar for Research at the University Carlos III in Madrid, the University Cergy-Pontoise in Paris and the Cass Business School in London since 2012.

He is the author of books and scientific publications in international and national journals. His main research interests include issues of Banking and Finance, Corporate Governance, ESG and sustainability features, Law and Finance, Law and Economics, microfinance issues; corporate finance and corporate histories of banks and industrial corporations. He is Professional Accountant since 1995 and Auditor since 1999 and is or was a member of the board of directors or of the board of auditors in Italian listed corporations or corporations belonging to listed groups and in corporations controlled by public bodies.

Laura Pellegrini (B.A. Economics of Institutions and Financial Markets, (Milan Catholic U.), Ph.D. in Economic Policy (Piacenza Catholic U.) She obtained National Qualification as Associate Professor of Banking and Finance in 2017. She is a Research Fellow, Lecturer B, University of Bari, Aldo Moro. She was Adjunct Professor of Corporate Finance, Faculty of Banking, Finance and Insurance Sciences, Catholic University of the Sacred Heart of Milan, where she teaches Corporate Finance. She is a Research Fellow of the Research Centre for Studies in Applied Economics (CSEA), Catholic University of the Sacred Heart, Milan and she was Post-doc Research Fellow in Corporate Finance (Catholic University), and Visiting Ph.D. student at the Warwick U. Economics Dept. (UK).

Her main research interest includes issues of Banking and Finance, Corporate Governance, Corporate Finance, Systemic risk and Sustainability features, Law and Finance and Law and Economics, Political Connections and Corruption, Stability and efficiency issues in financial intermediaries. She is the author of books and several articles, at international and national level.

Marco Seracini is a contract Professor of Corporate Governance and member of the Scientific Committee of Applied Economics Study Center (CSEA) at the Catholic University of the Sacred Heart of Milan and lecturer at the University of Florence. He is a certified chartered accountant, Chairman of the Board of Statutory Auditors and Standing Auditor of many leading companies, including listed companies, as well as director and auditor of companies, Public Entities and Foundations.

Abbreviations

AFME	Association for Financial Markets in Europe
AGENDA	United Nations (UN) 2030 Agenda for Sustainable Development
AUM	Assets Under Management
BICS	Bloomberg Industry Classification System
CAC	Contributo Ambientale CONAI
CAGR	Compounded Annual Growth Rate
CAPEX	Capital Expenditure
CAPM	Capital Asset Pricing Model
CAR	Cumulative Abnormal Return
CBI	Climate Bond Initiative
CE	Circular Economy
CEO	Chief Economic Officer
CERES	Coalition for Environmentally Responsible Economies
CFO	Chief Financial Officer
CFRF	Climate Financial Risk Forum
COE	Cost of Equity
COP	Conferences of Parties
CSR	Corporate Social Responsibility
EAI	Enhance Analytics Initiative
EBITDA	Earning Before Interest, Tax, Depreciation and Amortization
EBIT	Earning Before Interest and Tax
EIB	European Investment Bank
EIOPA	European Insurance and Occupational Pensions Authority
EPA	Environmental Protection Agency
EPS	Earning per Share

ESG	(Environmental, Social, Governance)
ESMA	European Securities and Market Authority
EU GBS	European Union Green Bond Standards
EU	European Union
FCA	Financial Conduct Authority
FFS	Forum per la Finanza Sostenibile—National Sustainable Investment Forum of Italy
FIR	National Sustainable Investment Forum of France
FNG	National Sustainable Investment Forum of Austria-Germany-Switzerland
FP	Firm' Profitability
GAAP	Generally Accepted Accounting Principles
GDP	Gross Domestic Product
GHG	Greenhouse Gas Emissions
GICS	Global Industry Classification Standard
GISR	Global Initiative for Sustainability Ratings
GRI	Global Reporting Initiative
GSIA	Global Sustainable Investment Alliance
HLEG	High Level Expert Group
ICMA	International Capital Market Association
IFRS	International Financial Reporting Standards
IIRC	International Integrated Reporting Framework
IMF	International Monetary Fund
INC	Intergovernmental Negotiating Committee
IPCC	The Intergovernmental Panel on Climate Change
IPSF	International Platform on Sustainable Finance
JSIF	Japan Sustainable Investment Forum
M&A	Merger and Acquisition
MIFID II	Markets in Financial Instruments Directive
MVA	Market Value Added
NDC	Nationally Determined Contributions
NFRD	Non-Financial Reporting Directive
NGOs	Non-Governmental Organizations
OECD	Organization for Economic Co-operation and Development
PLAN	Action Plan on Financing Sustainable Growth
PRI	Principles for Responsible Investment
R&D	Research and Development
RIAA	Responsible Investment Association Australasia
RIA Canada	Responsible Investment Association Canada
ROA	Return on Assets
ROE	Return on Equity
ROI	Return on Investments
ROS	Return on Sales

SASB	Sustainability Accounting Standards Board
SB	Società Benefit
SDG	Sustainable Development Goals;
SPAINSIF	National Sustainable Investment Forum of Spain
SRI	Sustainable Responsible Investment
STI FORUM	Science Technology and Innovation Forum
TCFD	Task Force on Climate-related Financial Disclosure
TEG	Technical Expert Group on Sustainable Finance
TFM	Technology Facilitation Mechanism
UKSIF	UK Sustainable Investment & Finance Association
UNEP	United Nations Environment Program
UN WCED	United Nations World Commission on Environment and Development
UNFCCC	United Nations Framework Convention on Climate Change
USSIF	United States Sustainable Investment Forum
VBDO	Dutch Association of Investors for Sustainable Development—National Sustainable Investment Forum of Netherlands
WEF	World Economic Forum

List of Figures

List of Tables

PART I

Climate Change challenge, CSR and ESG Issues: The State of the Art

CHAPTER 1

Introduction

Carlo Bellavite Pellegrini

During these last years, sustainability issues are becoming increasingly important, because everyone acknowledges sustainability is not only connected with environmental aspects or with corporate governance or with something else about financial and economic evidences, but it is strictly intertwined with our daily life and with mankind's future. It is not only a business's issue; it is a survival's one.

In 2020 the unpredictable and unexpected spreading all over the world of the pandemic disease Covid-19 strongly underpinned the importance of drawing reliable sustainable economic and financial paths for corporations and sustainability compliant legal and political contexts for nations. National governments and international institutions and authorities are effectively likely to be fully committed to engaging sustainability issues in

C. Bellavite Pellegrini (✉)
Department of Economic Policy, Catholic University of the Sacred Heart, Milan, Italy
e-mail: carlo.bellavite@unicatt.it

Research Centre of Applied Economics, Catholic University of the Sacred Heart, Milan, Italy

© The Author(s), under exclusive license to Springer Nature Switzerland AG 2022

C. B. Pellegrini et al. (eds.), *Climate Change Adaptation, Governance and New Issues of Value*, Palgrave Studies in Impact Finance,
https://doi.org/10.1007/978-3-030-90115-8_1

any sort of pandemic disease's recovery plans like Next Generation EU they are scheduling for the next years.

We have to admit that financial and economic scholars, practitioners, advisers, and managers were and still are only partially prepared to face these new challenges. Until not a long time ago an orthodox shareholders' vision (Friedman, 1970) was the only feasible doctrine allowed to have a global academic or managerial citizenship. Any kind of unorthodox belief, like a stakeholders' vision or even an "enlightened" shareholders' vision would have been banned in the global arena. Unorthodox views would have eventually jeopardized managerial or academic careers, being the juridical framework as well oriented in fostering an orthodox "*shareholder*" (Pistor, 2019).

Pope Francis recently mentioned it was originally unclear to him when he joined Aparecida's Conference (2007) why South American Church was devoting so much attention to Amazonia. He added to have realized how Amazonia did matter in ecclesial life, when he was becoming Pope, i.e., some years later. Moreover, he argued to have recently promoted, in late 2019, a Bishops' Synod, involving the eight Bishops' conferences of countries which insist within the borders of Amazonian Forest. From a historical perspective we have to acknowledge him to have worldwide promoted these issues since his 2015 Encyclical "Laudato sii," being the title of the papal document, the first sentence of the famous preach of Saint Francesco d'Assisi.

Eight centuries ago this latter raised substantial environmental issues within the medieval Western world, pathing the way for the development of a biblical and theological enquiry about creation and environment.

Amazonia matters for different reasons, declining sustainability issues according to a wide range of perspectives, like environment, community, natural resources, natural biodiversity, international relations. Amazonia encompasses all these items in just one concept about the physical, political, and even corporate future of mankind. From this point of view, Amazonia and its overwhelming biodiversity may represent an excellent metaphor, for corporate life, the polyhedral shapes of value, some of them still unexplored.

Notwithstanding all these boding well premises heading the path toward a more engaged sustainability perspectives of corporate life, I was personally taken by a very positive surprise on August 19th 2019 when 181 CEOs of outstanding American corporations[1] gathering at the Business Roundtable released a new Statement on the Purpose of a Corporation. This statement was finalized to commit themselves to lead their companies for the benefit of all stakeholders—customers, employees, suppliers, communities, and shareholders. This Statement represents a corporate revolution.

On the other side Reich (2021) asserts that corporate source responsibility may be seen as an attempt of the biggest corporation to avoid heaven fiscal burdens.

Like any revolution however we are entering in unknown fields. It is not only an issue about the purpose of the corporation, eventually underpinning an ontological distance from Milton Friedman's famous assumptions, but other issues, encompassing the metrics of value as well.

From one side we have to recognize that below zero interest rate and the huge amount of liquidity poured into the market by monetary authorities in order to challenge the financial crisis and in more recent times the pandemic disease are important ingredients in order to overcome the shareholder's vision in the short run.

On the other side, the above-mentioned statement discloses an innovative approach to these issues in the long run, simply because not sustainable businesses have no future. Sustainability is strictly connected with discontinuity about previous approaches in economics and business. Any sort of discontinuity juxtaposes great challenges and meaningful opportunities (Tamburi Investments Partners, 2017).

It is not easy to find a feasible mixture of these two concepts overlapping different aspects of corporate life, like business ventures, political choices, and technological opportunities (Filippetti, 2019).

This book tries to explore how we may detect some innovative measures of values in an extremely rapid evolving world. In order to achieve this outcome, we have to recognize how a multi-stakeholder world pushes us to consider a more composite definition of value. To the same extent economic and financial studies are accustomed to using in their inquiries some measure of financial or accounting returns, like

[1] Jamie Dimon, Chairman and Ceo of JPMorgan Chase & Co. and Chairman of the Business Roundtable declared: "Major employers are investing in their workers and communities because they know it is the only way to be successful over the long term. These modernized principles reflect the business community's unwavering commitment to continue to push for an economy that serves all Americans".

depending on variables. This approach recognizes the relative scarcity of financial capital in comparison with other forms of capital. Whenever in the future we are going to experience the relative scarcity of other typologies of capital, it is likely we have to find innovative measures and metrics for them, being the one we are using to measure financial capital, completely unfit.

The present volume is organized into two main sections including both the theoretical framework and the empirical evidence on climate change issues and corporate increasing commitment to overcome them and create resulting value-adding opportunities.

The first section is composed of six chapters, discussing the main topics of this book such as climate change adaptation, various benefits of investing in adaptation, and the sustainability payoffs in terms of added value to the firm. The second section presents some interesting analyses related to the connection between sustainability commitment and corporate financial performance.

In chapter two, Catizone describes self-consciousness and awareness as drivers behind corporations' evolution in recognizing their role within the ecosystem and capturing changes in such an ecosystem. In addition, the author highlights how knowledge and awareness are fundamental for a corporation in order to evolve by dynamically adjusting its business and strategies.

The third chapter is dedicated to one of the most relevant "new frontiers issues" of corporate governance. The chapter deals with the climate change challenge as well as the combined action of initial policymakers in defining the environmental objectives. These objectives are progressively more interconnected to the theme of stability.

In the fourth chapter, Pellegrini explains the Environmental, Social, Governance indexes (ESG) as innovative tools in measuring corporate sustainability and their impact on firms' financial and operating performance and cost of capital.

In chapter five, Montalbetti takes into consideration circular economy which may represent the premise of the ESG debate.

Section one ends with chapter six in which Cannas and Pellegrini review the literature on ESG issues, reporting the most important studies about ESG scores' effects on the cost of equity.

Throughout the second section, five empirical chapters are presented with the aim to investigate a potential correlation between a company's commitment to sustainability and its financial performance indicators.

In chapter seven, Bellavite Pellegrini, Caruso, and Seracini explore the Oil and Gas sector investigating the effect of ESG scores on cost of Equity and firm's profitability.

In the eighth chapter, Bellavite Pellegrini, Caruso, and Cifone empirically focus their research on the Automobiles and Part sector, enquiring how to assess the impact of the ESG scores on companies' performance indexes.

In chapter nine, Barbieri and Pellegrini propose an empirical case study on a very peculiar sector as Pharmaceutical one in order to verify whether a connection exists between ESG issues and their operational and financial performance.

Chapter ten is focused on an innovative financial instrument linked to sustainability. More specifically Camacci examines the stock market reaction to the announcement of corporate green bond issuance.

Finally, in chapter eleven Cannas proposes a survey on sustainability issues in Emerging Markets in order to investigate a connection between ESG scores and firms' performance. Some concluding remarks will draw a research agenda for future research in each chapter.

References

Filippetti, S. (2019). I signori del lusso. Sperling & Kupfer.

Friedman, M. (1970). A Friedman doctrine: The social responsibility of business is to increase its profits. *The New York times Magazine, 13*, 32–33.

Francis, P. (2015). Laudato si. Vatican Press, May, 24, w2.

Pistor, K. (2019). *The code of capital*. Princeton University Press.

Reich, R. B. (2020). The system: Who rigged it, how we fix it. Knopf Doubleday Publishing Group.

Tamburi Investments Partners. (2017). Prezzi & valori. L'Entreprise Value nell'era digitale. Borsa, private equity, M&A, premi, sconti, errori e prospettive. Class Editori.

CHAPTER 2

Self-Consciousness and Awareness as Adaptation Enablers

Massimo Catizone

Introduction

Large international corporations manage a vast amount of capital and assets. As a result, corporations are in a position to affect the environment and the value of natural capital. Considering the environment as a free and not finite resource is fundamentally incorrect. There is a growing consensus that any firm that is failing to implement the necessary adaptation strategies, consumes and erodes natural capital without an effective mitigation strategy, would do so at the expense of other firms and households. In this case, when remediation is possible, the remediation costs would have to be borne by other stakeholders.

Directing capital towards adaptation projects should be viewed as an opportunity for all stakeholders, not just corporations. As argued by the Global Commission on Adaptation investing in adaptation can provide

M. Catizone (✉)
Research Centre of Applied Economics, Catholic University of the Sacred Heart, Milan, Italy

© The Author(s), under exclusive license to Springer Nature Switzerland AG 2022

C. B. Pellegrini et al. (eds.), *Climate Change Adaptation, Governance and New Issues of Value*, Palgrave Studies in Impact Finance,
https://doi.org/10.1007/978-3-030-90115-8_2

a triple dividend: loss avoidance, economic benefits and social and environmental benefits. Not investing in adaptation should be considered as an active investment decision. The consequences of such a decision are difficult to predict and quantify, but it is almost certain that such consequences will negatively affect the value of stakeholders' capital over time. Early warning systems, for example, save lives and assets worth many times their cost.

Due to the role of corporations and the magnitude of the size of the assets that are under the influence of their choices and decisions, it appears legitimate to argue that corporations have a responsibility and, more importantly, an interest, in implementing business strategies capable of preserving natural capital, monitoring and mitigating, through constantly refined risk management strategies and solutions, the impact of climate change.

The question then arises as to if and how corporations can and should go about identifying and implementing sustainable business strategies. Also, the parameters and reference points to be used by corporations to assess whether or not their strategies are suitable to achieve their objectives, generate value for all stakeholders and mitigate climate change-related risks should be identified. In this respect, two macro-areas should be considered.

First, risk identification and disclosure. As of today, disclosure of climate-related risks is, especially for smaller firms, largely voluntary and as a result, available data is scarce and, in most cases, inadequate to enable investors to compare corporations from the same sector and assess relative resilience to climate-related shocks.

Second, the mission of the corporation. As the stakeholders' base and its expectations change over time, the mission of the corporation should predict and acknowledge such changes and adjust accordingly. For example, the potential benefits of digitalization and new technologies as enablers of a green transition and as tools for climate change risk mitigation should be fully exploited. It has been argued (Patel et al., 2010) that machine learning algorithms can be successfully used to predict climate change and in the context of conservation planning by locating habitats of wildlife and predicting future sites where wildlife would be likely to relocate based on scientifically backed climate change assumptions.

Old Demons and New Challenges: Sustainability and Beyond

Over the years, a number of scholars and institutions have attempted to define the concept of sustainability. The UN World Commission on Environment and Development (World Commission, 1987) defined sustainable development as a form of development "that meets the needs of the present without compromising the ability of future generations to meet their own needs". A sustainable development should therefore enable the biosphere and human civilization to coexist in the long term. Sustainable development is by no means a new challenge. For example, it has been argued (Harper, 2017) that the Romans built an interconnected, urbanized empire on the fringes of the tropics with tendrils creeping across the known world. In an unintended conspiracy with nature, the Romans created a disease ecology that unleashed the latent power of pathogen evolution. The Romans were soon engulfed by the overwhelming force of what we would today call emerging infectious diseases. The end of the Roman empire then is a story in which humanity and the environment cannot be separated, a stark reminder of the magnitude of the impact that nature and climate changes can have on our ways of life. A similar event, but on a much more reduced scale happened in 2016 and 2017 when public fears over the Zika virus eroded hotel tax collections across the state of Florida and other international markets. Other consequences included costs associated with treating birth defects among features and infants of US women with evidence of possible Zika Virus infection during pregnancies. The World Bank estimated that the economic costs of Ebola in West Africa due to disruptions in travel and trade exceeded half a billion dollars. An estimate published in Health Economics placed the cost of lost tourism revenue in Mexico during the 2009 Swine Flu outbreak at $2.8 billion. The full extent of the long-term impact on global GDP, on demographics and our ways of life of the 2020 coronavirus outbreak is yet to be determined. According to a report from OECD the scale of the estimated decline in the level of output is such that it is equivalent, in the absence of offsetting factors—to an annual decline in annual GDP growth of up to 2 percentage points for each month that the strict containment measures continue. Due to the size of the expected contraction in global output, there is a risk that—at least in the short to medium term—governments sustainability strategies may be relaxed. Because of the global economic recession and the imminent threat posed by it, accelerating economic recovery is viewed by many as a priority not compatible with the implementation of climate-related policies. It cannot be excluded that, due to the lockdowns and other containment measures imposed by many governments, virtually all countries, including those that have been less proactive in responding to the challenges arising from climate change, will meet their CO_2 reduction targets. Containment measures resulting in a reduced entitlement to enjoy certain rights that in most jurisdictions are perceived as fundamental are only acceptable for a limited period of time and in exceptional circumstances. Not surprisingly there is a consensus that these measures should not be regarded as sustainable long-term solutions for tackling climate change (Fig. 2.1).

The 2020 pandemic is emphasizing once again the magnitude of the challenge deriving from climate change, which has been eloquently defined—Bodansky et al. (2017)—as planetary in scope, and because of its potentially irreversible consequences intergenerational in its impact. Again, we witnessed the close relation between loss of biodiversity and

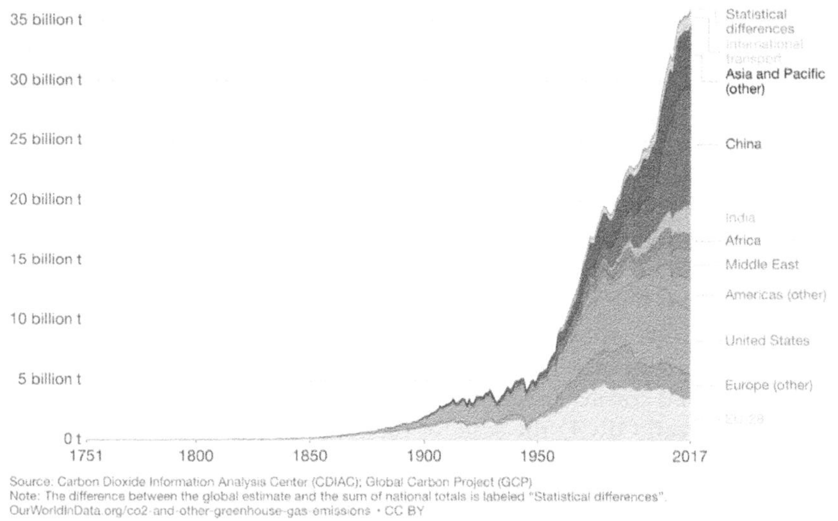

Fig. 2.1 Annual total CO_2 emission, by world region, 1751–2017 (*Note* The difference between the global estimate and the sum of national totals is labelled "Statistical differences". *Source* Carbon Dioxide Information Analysis Center [CDIAC]; Global Carbon Project [GCP])

disease outbreaks and that the recovery will require *ad hoc* solutions. In April 2020 the EU Technical Expert Group on Sustainable Finance issued a statement emphasizing that sustainable recovery from the COVID-19 pandemic requires the right tools which, according to such statement are to be found in the following EU documents: (a) Sustainable finance taxonomy, (b) Green Bond standards and (c) Paris aligned and climate transaction benchmarks. These tools encourage substantial contributions to environmental and social objectives, also from corporates and financial actors that are not yet fully aligned with environmental goals.

The pandemic seems to be advocating the need for a bold shift towards integrated global macro strategies, a new role for governments and law makers, significant changes in how we consume goods, interact with our peers, protect our personal data and possibly even a reconfiguration of the ranking of our basic rights. Put simply, the financial and medical support that governments are providing during the pandemic may ultimately be subject to conditions intended to reduce the risks of a second wave of infections or the spread of a new virus. The global GDP contraction caused by the pandemic was a reminder of the importance of social capital and, as argued by Kwon and Arenius (2010), of social entrepreneurship and that the current equilibrium between government, markets and social capital may benefit from some rebalancing.

These are not trivial challenges and it should not be taken for granted that commonly accepted principles, concepts and solutions can be relied upon to overcome such challenges. History has shown us that crisis are defining moments leading to the emergence of new leaders and the acceleration of innovation.

Arguably, a modern concept of sustainability should go beyond the limits of the mere coexistence of human civilization and nature and encompass the relationships between nations. There is a consensus that environmental factors are often important drivers behind conflicts. Nations have often fought to assert or resist control over raw materials, energy supplies, river, land, sea passages and other key environmental resources. Such conflicts are likely to increase as these resources become scarcer, the human population becomes larger and competition for resources increases. Environmental stress can thus be an important part of the web of causalities associated with any conflict and can in some cases be catalytic. Extreme weather events and pandemics whether or not anthropogenic can trigger or accelerate migration flows and cause international political instability.

Is the GDP of nations with a greener and more circular economy going to be less vulnerable to international natural-capital-dependent tensions and conflicts? Would a reduced reliance on finite resources reduce the probability of conflicts? It has been pointed out that better environmental policies reduce the likelihood of conflicts and that a circular economy, almost by construction, would be conducive to peace and prosperity (Behrens, 2016; Caruso et al., 2016).

These questions are extremely complex, and the answers are going to be very articulated, but there is growing evidence suggesting that countries that are less dependent on oil imports would be less affected by large scale conflicts involving oil-producing countries. When in January 2020 US President Donald Trump announced the death of Qassem Solimani, oil prices increased more than 3%, but soon thereafter reverted to previous levels. The initial price movement was driven by a number of factors including investors' concerns on the magnitude of Iran's possible retaliation. Concerns were subsequently eased by a number of political and financial factors. The relatively limited reliance of the US economy on oil imports and the ability of US corporations to access alternative sources of energy supply and deliver alternative innovative and efficient energy production solutions were among the main reasons that lead to a rapid price reversion.

This seems to suggest that an economy that is less reliant on the import of fossil fuels and is able to derive its energy from sustainable, renewable and eco-friendly sources would be less vulnerable to the consequences that international tensions and conflicts may have on the prices of fossil fuels. Furthermore, commercial tensions may arise between neighbouring countries with significantly different regulations and limits applying to CO_2 emissions. The president of the European Commission in her 2020 speech at Davos warned China that a Carbon Import Tax may be applied to imports of Chinese goods should China fail to implement a credible CO_2 emission reduction strategy. More importantly, it may be highlighted that an economy that is less reliant on the supply of finite energy, greener, more circular more sustainable and characterized by a good governance system, would be more likely to be conducive to peace, stability and value creation than an economy that remains more fossil fuel dependent and less willing to implement strategies capable of mitigating the impact of climate change (Caruso et al., 2016). Typically, resources abundance and resource dependence (Lashitew & Werker, 2020) have a different effect on development. Resource abundance tends to have a positive direct impact on development. Resource dependence has a stronger indirect impact on human capital development. It has also been observed (San-Akca et al., 2020) that there is a link between energy policies and conflicts intervention. Furthermore, it has been argued (Maniruzzaman & Al-Saleem, 2017) that at a local level suboptimal governance mechanisms and uncertainty surrounding the legal framework applying to the exploitation of fossil fuels is a source of legal risk, internal friction and a factor negatively affecting sustainable development.

What is the taxonomy of a sustainable economy? What constitutes a sustainable fishing policy? What is a sustainable water management strategy? What is sustainable deforestation? What is a sustainable immigration policy? What does a sustainable banking industry look like?

Defining sustainability is a largely theoretical exercise based on mostly arbitrary assumptions. There is a growing body of literature suggesting that it should not be taken for granted that sustainability is the correct tool for calibrating our response to the challenges posed by climate change. Some scholars are going as far as arguing that the objective of environmental policies should not be sustainability, but resilience. They state that defining sustainability is a difficult, largely unachievable and futile goal. Furthermore, it has been observed that one of the main weaknesses of sustainability is that it erroneously assumes that "anthropogenic changes are non-transformative and hence (generally) reversible". There is now a consensus that at least some of the anthropogenic changes to the environment are not reversible. Some scholars are therefore arguing that the priority of any response to the challenges posed by climate change should be the creation of an ecosystem that is able to absorb the shocks deriving from the inevitable and irreversible climate changes. It can be underlined that the merits of a resilient ecosystem are unquestionable. Any solution to the challenges posed by climate change should therefore ultimately be aimed at creating a sustainable environment characterized by a durable equilibrium among all stakeholders and their interests.

Sustainability and resilience do not appear to be mutually exclusive. It is difficult to imagine a sustainable ecosystem that has zero or limited shock absorption capacity. It could therefore be argued that resilience is an essential pre-condition for sustainability. It may also be argued that the importance of finding a consensus on the definition of sustainability, sustainable investment, sustainable economy, sustainable governance, etc., should not be overemphasized.

What seems to deserve more attention is how the transition from the *status quo* to a more resilient and stable long-term equilibrium between stakeholders interest can be achieved.

The main common thread between sustainability theories and resilience theories are the need for adaptation. In order to respond to the climate change challenges, virtually any ecosystem or part of such ecosystem need to adapt and evolve. As a consequence, it appears legitimate to envision that the currently prevailing definition of corporate mission, agency theory and a number of other commonly accepted economic theories and principles may have to be reconsidered or adjusted considering the new findings on the relevance of adaptation, the constitutional elements of corporations, their priorities and purposes.

The Conscious Corporation

In law, a legal person is any person or entity that can do the things an everyday person can usually do in law such as enter into contracts, sue and be sued, own property, etc. Smith (1928) argues that to be a legal person is to be the subject of rights and duties. To confer legal rights or to impose legal duties, therefore, is to confer legal personality.

Some of the main defining features of a natural person are self-consciousness and consciousness. Consciousness generally involves the perception of the physical environment based on knowledge and experience. Self-consciousness involves the perception of being physically distinct from, but part of, a surrounding ecosystem. Consciousness and self-consciousness would therefore enable a corporation to define its mission, its role within a particular ecosystem and adjust them as the objectives and priorities of the ecosystem change.

Self-consciousness would enable the corporation to engage with all stakeholders with the purpose of creating consensus and demonstrating the alignment of its interest with the interest of the ecosystem of which it is an active stakeholder. Engagement is essential as it is the continuous process that allows corporations to interact with their stakeholders. In the absence of such a process corporates become disconnected from their stakeholders and lose value. The purpose of any engagement between the corporation and its stakeholders should have multiple dimensions, but it should be primarily directed at demonstrating how the corporation is able to serve and create value and remain relevant for the entire stakeholder's base in the long term. A purposeful engagement is not possible without self-consciousness. Without knowledge and awareness, the corporation would lack the necessary tools to recognize its role within the ecosystem, capture changes in such ecosystem, in stakeholders aspirations and expectations and ultimately would be unable to evolve by dynamically adjusting its business and strategies. In this context, corporate governance plays an essential role. It is thanks to its corporate governance mechanisms that a corporation can develop its self-consciousness and engage with purpose and knowledge with its stakeholder's base.

Corporate consciousness enhanced by knowledge and purpose appears to be a very powerful and effective key to unlocking the complexities and managing the challenges associated with climate change.

Corporations that will be able to develop self-consciousness and engage with purpose will be ideally positioned to evolve, create value for all stakeholders, adapt and attract fresh finance flows.

The ability to adapt existing business models or existing products with the view to transforming the challenges deriving from climate change into business opportunities is critical. In Europe, for example, a growing group of banks has started granting mortgage loans secured against energy-efficient properties. Furthermore, the same group of banks is arguing that loans backed by energy-efficient properties have a lower credit risk compared to loans backed by traditional properties. The rationale for this argument is twofold. First, as the running costs for energy-efficient properties are lower, the borrowers would be more resilient, should any event with a negative impact on their income occur. Second, the price of energy-efficient properties tends to be more stable compared to traditional properties. The ultimate objective of this exercise is to obtain from the regulators a privileged regulatory treatment, in the form of a lower risk weight, for mortgage loans secured against energy-efficient properties. An initial temporary pilot project was successfully launched in February 2020 by the Hungarian Central Bank. Should a favourable view be taken at the European level, this could become an example of conscious corporations actively engaging with the climate change ecosystem for the benefit of all its stakeholders. One of the main obstacles to this scheme being rolled out across Europe is the lack of reliable data on the performance of loans backed by energy-efficient properties.

One of the main challenges that the modern conscious corporation is facing is its strategic positioning relative to the ecosystem within which it is operating. To a large extent, in the case of the ecosystem required to manage climate change, which we will refer to going forward as the Climate Change Ecosystem, the challenge derives primarily from the complexity of the issues at stake, the large quantity of stakeholders involved and the variety of interests, objectives and priorities of such stakeholders.

Governing the Climate Change Ecosystem is therefore essential in order to enable the conscious corporation, but also the other stakeholders, to function and contribute to the pursuit of the objectives of the ecosystem.

In the following paragraphs, we will discuss in more details the crucial role of disclosure in the management of the challenges deriving from climate change, considering some approaches that could be suitable to manage its complexities.

Managing Complexity in the Climate Change Ecosystem

Governing the Climate Change Ecosystem is a challenging exercise. This is mainly due to the inherent complexities of the task, insufficient scientific knowledge, varying stakeholders' capacity to address climate change challenges, but also to some institutional inertia Meadowcroft (2009).

There is no hard evidence suggesting that in order to govern such ecosystem a particular form of governance would be more effective than others. Given the complexity of the task and the magnitude of the repercussions that the wrong choice may have on our ability to manage climate change, it appears preferable to rely on techniques and solutions commonly used to manage complex systems. A subsequent adjustment may be required or advisable in order to accommodate the specificities of the Climate Change Ecosystem, its objectives and priorities.

The choice should therefore be driven by a simple and pragmatic analysis. The most suitable governance model is therefore the model that is more likely to successfully address the objectives and priorities of the ecosystem it is intended to govern. Any approach adopted to govern an ecosystem should therefore enable innovation, high performance and most importantly system adaptability. Scientific innovation is essential to identify and implement the technical solutions underpinning a sustainable and climate change resilient economy. Without scientific innovation, the impact of capital would most likely remain below potential. Capital is an essential enabler of innovation as without intelligent capital, scientific innovation is unlikely to progress and financial products may not be able to unlock the opportunities embedded in climate change. High performance is essential as all stakeholders should be able to communicate and interact with the view to achieving the best possible results within the shortest timeframe. And here is where adaptability, one of the key components of the ecosystem, comes into play. As we have pointed out before, the Climate Change Ecosystem is dynamic as it must react to new findings, new events, new knowledge and climate change itself, which, to a large extent is inevitable. Like the conscious corporation, the ecosystem should be able to recognize and acknowledge changes in environmental conditions, in stakeholders' priorities and objectives and adapt quickly in order to remain high performing and therefore able to serve its purpose over time.

Modularity governance and nodal governance-based solutions are often used to manage large and complex systems. Some of such solutions may be used to govern the Climate Change Ecosystem as they develop from the acknowledgement that governance is characterized by a plurality of actors forming more or less interconnected governance networks (Burris et al., 2005) argue that governance members of an ecosystem develop forms of governance as a strategic adaptation to complexity. They highlight that governance in complex systems is organized in nodes—institutions with a set of technologies, mentalities and resources—that mobilize the knowledge and capacity of members to manage the course of events. Nodes are points on networks and networks are a prime means through which nodes exert influence. This governance framework appears to be suitable to govern the Climate Change Ecosystem as each node may represent one group of stakeholders. Modularity is also used to manage complex systems. There is however some evidence suggesting that excessive levels of modularity may jeopardize system adaptability.

In light of the nature and ultimate objective of the Climate Change Ecosystem, it can be stated that any governance solution which may in any way negatively affect the ability of the ecosystem to adapt would not be the optimal choice. Therefore, should a compromise between performance and adaptability be inevitable, adaptability should prevail at the expense of performance. Ability to adapt or the lack of adaptation skills (Guay et al., 2015) are not only pre-condition for firm performance, but also CEOs longevity hence stability.

Consequently, we remain agnostic with respect to the most suitable form of governance, provided that, the governance mechanisms chosen by the ecosystem enable it to adapt swiftly. The ecosystem should adapt to the changing needs of its stakeholders, but stakeholders should also adapt and adjust the way in which they conduct their business in order to fit into the ecosystem and accommodate the objectives and requirements of the ecosystem and other stakeholders. Adaptation is therefore an essential condition for the Climate Change Ecosystem to prosper, but also a condition for individual organizations to become and remain part of the ecosystem.

The challenges that the Climate Change Ecosystem is facing are not new, however, the ecosystem itself is, in many respects, in its infancy. The ecosystem, through its governance mechanisms, should learn how to identify and manage such challenges. Failure to do so could negatively affect the efficiency and the growth pattern of the ecosystem. In this respect there are three main areas of risk that tend to be underestimated and that deserve to be monitored closely: (α) excessive regulatory fragmentation; (β) failure to assess and quantify correctly the ramifications of the measures that are implemented to respond to specific policy requirements and, last but not least, (γ) the risk that incentives presented as forms of legitimate support for sustainable economic activities constitute in reality unlawful state aid or other forms of subsidies that could distort competition within the ecosystem.

Excessive regulatory fragmentation. The rules governing the Climate Change Ecosystem derive from different organizations and a variety of initiatives. Some are purely domestic, others are intended to apply regionally, like the European Commission's Action Plan on Financing Sustainable Growth or internationally, like the United Nations initiatives. Some decree of regulatory and supervisory fragmentation is inevitable and to some extent potentially beneficial for an ecosystem as complex as the Climate Change Ecosystem. However, excessive supervisory and regulatory fragmentation and a lack of harmonization should be avoided as they could lead to operational inefficiencies, duplication of costs and ultimately may affect the ability of the ecosystem to function efficiently.

The possibility that the repercussions and consequences on the ecosystem of a specific economic activity may not be fully appreciated is also a risk. In order to mitigate this risk, the Technical Expert Group proposed a "do no harm" test. In order to determine the environmental sustainability of economic activity, such activity should meet at least one policy objective, but also it should not have a direct or indirect negative impact on other policy objectives. This appears to be a relatively straightforward test, but meeting its requirements is expected to be excruciatingly difficult.

It can be argued that incentives aimed at accelerating, especially in the early phases, the growth of an ecosystem are beneficial. Furthermore, providing supports and incentives is an important component of the role of governments and lawmakers in the Climate Change Ecosystem. However, the ultimate objective is for the ecosystem to be economically and financially viable, sustainable, resilient, fair and inclusive. Therefore, any form of incentive contravening competition rules or providing undue benefits to a particular group of stakeholders at the expenses of the others should be inhibited. It should be recognized that drawing a line between genuine green incentives and sophisticated forms of state aid is a complex exercise, which is exactly why it is argued that this is a risk that deserves ongoing monitoring.

An additional level of complexity derives from the fact that there is often an overlap between sectors that are considered strategic by governments, such as energy, construction, infrastructure or the housing market and sectors that are the subject matter of environmental social and governance policies. As a result, there appears to be a risk that specific measures may not be entirely implemented to support a transition to a more sustainable and resilient economy, but simply to advance a political agenda. ADAPTATION IN THE CLIMATE CHANGE ECOSYSTEM: RECONSIDERING THE CORPORATE MISSION AND THE DISCLOSURE OF CLIMATE CHANGE RELATED RISKS.

Since 2015, most of the countries that have signed up to the Paris Agreement have been adapting their existing legal and regulatory frameworks in order to encourage new capital to be deployed against initiatives that are instrumental to the achievement of its the main long-term purposes objectives of the Paris Agreement.

The Paris Agreement is intended to enhance and implement the objectives of the United Nations Framework Convention on Climate Change, adopted in New York on 9 May 1992, by, according to Article 2 thereof: (a) holding the increase in global average temperature to well below 2C above pre-industrial levels and pursuing efforts to limit the temperature increase to 1.5C above pre-industrial levels (...); (b) increasing the ability to adapt to the adverse impact of climate change (...); (c) making finance flows consistent with a pathway towards low greenhouse gas emissions and climate-resilient development. The Paris Agreement is generally regarded as a turning point in the history of climate change and climate governance. In this respect, A. Savaresi (2016) argues that the Paris Agreement marks the beginning of a new season for climate governance as "*it adopts a collective long-term goal on climate change mitigation to be supported by efforts from all parties (…) and for the first time recognizes the role of non-state actors in addressing climate change, and raises the profile of adaptation, albeit only incrementally*".

In order to be successful, the Climate Change Ecosystem should succeed in directing fresh capital towards projects supporting a more sustainable and resilient economy. It should be recognized that capital can facilitate and accelerate adaptation. However, it should also be recognized that the need for adaptation is more fundamental for the corporation than just a tool for attracting new capital. The Climate Change Environment is dynamic in nature and evolving constantly. The failure of the corporation to adapt and to become more resilient to the shocks deriving from climate change could eventually lead to the downfall of the corporation itself. Adaptation, intended as an innovation driven by a chain of destructive events (Schumpeter, 1976), is therefore a condition for attracting fresh capital, but more importantly, for business continuity. As eloquently observed by Charles Darwin (1859), "*it is not the strongest of the species that survives, not the most intelligent that survives. It is the one that is the most adaptable to change*".

Adapting Is not an Option but a Condition for Survival

Failing to adapt would threaten a corporation competitiveness first and potentially its existence in the long term.

Investors are reconsidering their assessment of climate-related risks and are gradually demanding higher rewards for taking climate risk. The more sophisticated investors are already recalibrating their assumptions across all asset classes including those that historically have been perceived as less risky. For example, investors in US municipal bonds are now expecting to be reworded for the risk of suffering losses deriving from increasingly frequent extreme weather events such as floods and earthquakes and their negative impact on the ability of cities to fully and timely repay their debts.

Directing capital towards adaptation projects should be viewed not only as a challenge but also as an opportunity for all stakeholders, not just corporations.

Not investing in adaptation should be considered as an active business or investment decision. The consequences of such decisions are difficult to predict and quantify, but early evidence suggests that these choices will affect the value of stakeholders' capital over time. Early warning systems, climate-resilient infrastructure, for example, save lives and assets worth several times their cost. The cost and benefit analysis conducted in respect of some of the most eminent climate-resilient infrastructure projects seems to support this view. The Netherlands, with the closing off of the Southern Sea in 1901 and the Delta Works commenced in 1953 offer an interesting case study in flood risk and water management. The North Sea flood of 1953 alone claimed more than 1800 lives. Following the 1953 dramatic events, the Dutch Government built a network of dykes known as The Delta Project or Delta Works. At inception, the estimated cost of the works amounted to 7.3% of the then GDP of the Netherlands and not surprisingly justifying and authorizing the Delta Works was no trivial challenge for the government and policymakers. A century later, CBP Netherlands Bureau for Economic Policy Analysis published a review of the cost and benefit analysis conducted at the time in a report entitled Cost–benefit analysis for flood risk management and water governance in the Netherlands: an overview of one century (2017). According to the report, both costs and benefits were underestimated. Interestingly, the report also outlines the main technical and regulatory challenges posed by the project which by no means should be considered specific to water management, but should be regarded as obstacles to be overcome by any climate change related physical risk mitigation project. The report emphasizes, as confirmed by OECD (2011) that water policy in many countries is ineffective due to fragmentation of tasks between different parts of government and lack of technical and scientific capacity. The outcome of the report seems to support the thesis according to which choosing not to invest in climate change risk mitigation is not a real option. Furthermore, the report also emphasizes the importance of governance capable of managing complexity and that scientific and financial innovation are essential to successfully manage climate change related challenges and opportunities.

Due to the role of corporations and the magnitude of the size of the assets that are under the influence of their choices and decisions, it appears legitimate to argue that corporations have a responsibility and, more importantly, an interest, in implementing business strategies capable of preserving natural capital, monitoring and mitigating, through constantly refined risk management strategies and solutions, the impact of climate change. Such strategies should also allow for corporates to adapt and make strategic and tactical decisions in response to the likely consequences of possible climate changes.

Climate changes are not easy to predict with accuracy and firms should assess the impact of climate change on their business based on risk pathways embedding the effects that climate change related risks are likely to have on their profitability and value over a time horizon that is consistent with the timeframe within which climate changes may have tangible effects. Different stakeholders have different time horizons and a long-term approach, especially among investors, is not widespread. As a result, corporations may be under pressure to pay dividends, institutional investors may be under pressure to outperform their peers or a benchmark in the short term and financial institutions may find long term lending penalizing from a regulatory capital management point of view. Therefore, there is a risk that in the absence of systemic governance capable of implementing incentives, rewards and sanctions creating an alignment of interest among stakeholders, short-termism may prevent or slow down the growth of sustainable finance initiatives.

The question then arises as to if and how corporations can enhance their ability to absorb shocks deriving from climate and identify and implement sustainable business strategies and what parameters and reference points should be used by corporations to assess whether or not their strategies are suitable to achieve their objectives, generate value for all stakeholders.

Furthermore, corporations should allow investors and other stakeholders to take an informed, data-based view on the effectiveness of the actions taken in order to take advantage of the opportunities and mitigate the risks deriving from climate change. In this respect, two macro-areas should be considered: risk identification and disclosure and corporate mission. As regards corporate mission, in the following paragraphs, we will discuss the merits of an adjustment to the traditional agency theory based on a concept of corporate mission that reflects the need for the conscious corporation to be a part of the Climate Change Ecosystem. In the next paragraph, we will discuss risk identification and disclosure.

Risk Identification and Disclosure

As of today, disclosure of climate-related risks is not compulsory for all market participants and as a result, available data is scarce and, in most cases, inadequate to enable investors to compare corporates from the same sector and assess relative resilience to climate-related shocks. Disclosure requirements should be determined with a view to meeting investors requirements, facilitating financial supervision and measuring the resilience of the economy against climate change related shocks. Investors requirements and expectations in respect of disclosure will largely depend on the usage that investors intend to make of the environmental, social and governance data and information. Such usage and the required level of granularity of such disclosure will ultimately depend on the investment strategy of each investor. For example, investors that use ESG data for screening purposes only are likely to be content with a lower level of detail and data granularity than investors that fully integrate ESG data in their investment analysis. Active investors will expect full quantitative disclosure, but will also require full visibility and direct participation in corporate governance matters.

In this respect, a survey conducted by Amel-Zadeh and Serafeim in 2018 provides some insight on why and how investors use reported ESG information. The outcome of the report suggests that the majority of investors use ESG because it is financially material to investment performance. A smaller cluster of investors considers ESG information because of growing client demand or formal clients' mandates. A residual group considers ESG information because they see it as their ethical responsibility. Finally, a small group of US investors do not use ESG information as they believe that this would violate their fiduciary duties towards their stakeholders. Climate change risks should be treated, when possible, like any other risk to which a corporation is exposed. Disclosure requirements should be determined in order to provide investors with a fair representation of the objectives, or lack of them, achieved by the disclosing entity. Disclosure should also enable investors to determine the relationship between financial data and environmental, social and governance data.

There is a growing body of literature research suggesting that the biggest challenge to using ESG data for investment decisions is data noise and the lack of comparability of reported information across firms. Lack of reporting standards, costs of gathering and analysing ESG data is also often regarded as a major obstacle. One of the major negative consequences of the current status of the climate change risk disclosure is that as of today there is no well-defined and commonly agreed materiality test for disclosure of ESG factors. In many cases, it is the disclosing entity that independently decides what is material and should therefore be disclosed. A concept of materiality should be recognized and ideally arrived at based on the defining features of the "E", of the "S" and of the "G". More progress appears to have been made on the "E" and on the "G" than on the "S".

Defining the perimeter of the "S" disclosure seems to represent a growing challenge, due to its complexity and urgency.

The urgency arises from the fact that the social ramifications of investments appear to be particularly relevant for an increasingly large and active group of investors: women and millennials. Failing to address their expectations to learn how their capital contributes to human rights protection or how they contribute to the advancement of social objectives may result in significant amounts of capital being directed away from sustainable investments, or more precisely, investments that could be qualified as sustainable if a satisfactory disclosure of the "S" factors was provided. This task appears to be challenging for two main reasons. First, save for a growing, but yet relatively small, group of investors, mainstream investors have limited interest in "S" factors. Their interest tends to materialize only when such factors lead to short term costs that are easy to calculate. Such costs are most likely to occur when mismanagement of social issues result in damage to brand reputation, lawsuits, fines, workplace shutdowns or consumer protests. In this respect, the analysis of operative provisions in international treaties seems to provide a marginally more comforting scenario. Clauses relating to human rights are now a common feature in international treaties and are more likely to appear in international treaties than clauses on sustainable development. For example Articles 72 and 73 of the EU-Cariforum agreement states that the Parties are required to act in accordance with core labour standards, not to operate their investments in a manner that circumvent international labour or environmental obligations and to ensure that foreign direct investment is not encouraged by lowering domestic environmental, labour occupational health and safety legislation and standards or by relaxing core labour standards or laws aimed at protecting and promoting cultural diversity.

The complexity of assessing the impact of the "S" and more generally of human capital arises from a number of factors. It is exceedingly difficult to measure the impact of upgrading facility safety or regulating hours of work. In this respect, however, a growing body of research suggests how intangibles contribute more and more to corporate value creation. For example, firms that do not treat their human resources well typically suffer higher turnover affecting productivity, trust and innovation. This is expected to negatively affect the value of the corporation. More generally, while there is a consensus that human capital plays an important role in economic development and that when assessing human capital, the focus should not be on school attainment, but on cognitive skills and school quality, a full consensus on an approach for measuring the actual impact of human capital and social capital on economic growth is yet to be reached.

Finally, there is a consensus that in order for disclosure to be effective, its focus should shift. Disclosure should not simply be a factual description of the measures and policies implemented by the corporation in order to address climate change risks. This is an initial, and useful step, but by no means sufficient to enable investors to draw any meaningful conclusion. Therefore, in order for disclosure to be relevant, it should be focused on the actual objectives achieved by the specific measures and policies implemented.

In Europe, the Task Force on Climate-related Financial Disclosure has made progress in shaping a standard for voluntary disclosures by businesses and well over 900 public and private sector organizations have signed up to support it. The quality of the data is improving and is moving towards comparability. In the early 90s no more than 20 companies disclosed ESG data. Today more than 10,000 companies disclose non-financial risks.

With respect to risk identification, most corporates are disclosing macro risk areas: climate-related transition risk and physical risk. Transition risk arises from climate-related events that may damage corporate assets, infrastructure, the supply chain, or the natural environment in which the corporate operates. Transition risk arises from the shift to a carbon-neutral economy. This may impact asset values, corporate valuations, energy prices, the corporation customers' ability to honour their debts. Transition risk may arise from a variety of different factors including technological innovation, consumers tastes and preferences and legal and regulatory changes. The impact of stranded assets on company valuations is a practical illustration of how transition risk may affect the market value of the assets and consequently of the companies that extract, distribute or whose business is highly dependent on fossil fuels. Stranded assets consist of reserves that should in principle remain unutilized in order to ensure that global temperature reduction targets are met. Typically, assets become stranded as a consequence of regulatory changes, customers demand or innovation. Regulations imposing restrictions on the usage of fossil fuels may be implemented, customers may opt for greener sources of energy and innovation may make alternative and greener sources of energy more readily available. Impairments applied or to be applied to the market value of such reserves would have a direct impact on the value of the corporation. Predicting the magnitude of such impact requires taking a view on factors that are by nature difficult to predict and quantify, such as the timeframe within which a new regulatory framework may be phased in or how long would it take for a reliable alternative to fossil fuels to become commonly available. Identifying climate change related risks and facilitating such identification by way of enhanced disclosure does not appear to be sufficient to mitigate the impact of such risks. Knowledge should be enhanced and governance mechanisms capable of maximizing the usage and the impact of the knowledge contained in the various nodes of the ecosystem should be identified. Traditional risk management solutions should be adjusted accordingly.

Transition risk and physical risks consist of many interrelated moving parts and any risk management solution is often the result of a strategic compromise. For example, continued emissions will lead to rising temperatures that increase physical risk, but limiting these impacts require substantial emissions reductions that increase transition risk. Corporations should recognize that disclosing non-financial risks is an opportunity to demonstrate their understanding of such risks and facilitate the assessment of their business vulnerability to such risks by third parties. Transparency should also lead to a greater level of constructive engagement with all stakeholders. It has been observed that there is a positive association between ESG disclosure level and firm value, suggesting that improved transparency and accountability and enhanced stakeholders trust play a role in boosting firm value. The value of a corporation is somewhat subjective and, as discussed in more detail in the following chapters, different investors may assign a different value to different assets (especially intangible assets). Furthermore, performance and value can be measured in many different ways. In this respect it should not be taken for granted that the existing and commonly accepted measures of performance are suitable to capture what is relevant for all stakeholders or that the non-financial statements and the level of disclosure of physical risk and transition risk, which as of today remains largely discretionary, satisfy all the stakeholders.

Traditional risk management tools and skills may not be sufficient to assess transition and physical risks. Credit experts alone are unable to identify transition risk pathways and sustainability experts may not be able to conduct a comprehensive assessment of the impact of specific transition risk pathways on the creditworthiness of a corporation or of its customers. Corporations will have to enhance internal governance mechanisms to ensure that the usage of internal resources is optimized. Credit rating agencies are gradually starting to incorporate ESG valuations in their credit assessment underpinning both fundamental ratings and structured finance ratings. As of today, credit ratings do not yet incorporate a full qualitative and quantitative assessment of climate-related risks. It is however expected that rating agencies' involvement in this space will contribute to improving reporting standards and enhancing the general understanding of the impact on credit quality of climate-related risks.

Agency Model and Corporate Mission

The growing sophistication of the ecosystem in which corporates operate and the consequent evolution of the concept of corporate mission, constitute an additional layer of complexity. It has been observed that corporates are becoming accountable to a larger and more diverse stakeholders' base (Johnston 2008). Different stakeholders have different objectives. More importantly, the stakeholders' base is dynamic. Over time, as technology improves, scientific knowledge progresses and the way in which natural resources can be exploited evolves, and energy produced and stored, the stakeholders base and its expectations change. There are multiple ramifications to this complexity. It appears legitimate to suggest that certain aspects of the traditional agency model, which is often used to understand and manage conflicts between management and risk-takers should be reconsidered and expanded so as to capture within the definition of risk-takers, not only shareholders but all the stakeholders including those that are accidental risk-takers and whose interests are directly or indirectly affected by the business decisions of the corporation. Corporations should therefore acknowledge an implicit fiduciary duty towards a broader and fluid stakeholders base. Discharging a fiduciary duty towards a fluid stakeholders base is no trivial task as the stakeholders may change or the priorities of such stakeholders may change. In this context, in order to generate value, the corporation should be able to define its role within the ecosystem in which it operates independently and autonomously and based on overarching objectives. In order to do so efficiently, the corporation should also be able to understand the impact of its choices and actions on the ecosystem itself, which is another reason why it is pointed out that self-consciousness is a fundamental pre-requisite for a corporation that wishes to be an active and value-adding member of the Climate Change Ecosystem. The fiduciary duty should therefore be towards the ecosystem itself, its principles, its objectives and its participants, to the extent that their interests and priorities are aligned with those of the ecosystem.

Senior management and key employee's remuneration mechanisms play an important role in this respect. Compensation and its main components, including base compensation and variable compensation both corporate-performance linked and individual-performance linked should remain unaltered as they facilitate the alignment of interest between management and stakeholders but could also be used as a tool—mainly by deferring part of the compensation—to discourage short-termism on the part of the management. Due to the different layers of interests affecting the shareholders' base directly and indirectly some scholars (Frentrop, 2012) have proposed a more nuanced approach to the agency theory based on the assumption two agency relationships exist. The first one is the traditional one, where the shareholder is the principal and the company director is the agent. In the second agency relationship, the principal is the ultimate beneficiary of an institutional investor and the agent is the portfolio manager.

Directors and senior management involvement with the business is limited to the term of their tenure and should not be surprising that directors and management often tend to favour strategies leading to crystallization of profits during the term of their tenure at the expenses of long termism. This could lead to excessive risk-taking or moral hazard which could potentially erode value in the long term. With respect to the relation between bank failures and corporate governance, Berger et al. (2016) argue that defaults are strongly influenced by a bank ownership structure: high shareholding of lower level management such as vice presidents, increase default risk significantly. In contrast, shareholdings of outside directors and chief officers (managers with a "chief officer" position, such as the CEO, CFO, etc.) do not have a direct impact on the probability of failure. These findings suggest that high stakes in the bank induce lower level management to take high risks due to moral hazard incentives, which may eventually result in bank default.

The latest global financial crisis has undermined the traditionally accepted concepts of corporate mission and made scholars and practitioners more interested in identifying alternatives.

However, predominant economic theories suggest that managers should priorities the interests of the firm equity holders (Friedman, 1962). This objective is typically achieved by maximizing the present value of a firm's future cash flow. Some scholars have observed (Mayer, 2018) that if the attainment of profits does not come from their pursuit but is the product of some other purpose then the achievement of that purpose requires the management of other sources of capital such as natural, human and social as financial capital. From a legal point of view, it has been argued that directors are entitled to put other interests above profit maximization, provided that when deciding to do, so they are not conflicted and they act in what they believe to be the best interest of the firm. From a financial point of view, the position is more nuanced. It remains to be demonstrated if an alleged sustainable investment strategy can reduce the present value of future cash flows and generate monetary value at the same time. It has been argued that the answer to the question is affirmative when the demand exceeds supply. If supply and demands conditions are not favourable, engaging in socially responsible activities can actually reduce the market value of a firm. This finding does not solve the problem, especially because the impact of supply and demands dynamics are particularly relevant in the context of sustainable and adaptation investing as the amount of capital that is available to be deployed to support sustainable investments is relatively large compared to the size of the permitted eligible investments.

What is the legal framework that is most suitable to allow corporations to adapt and evolve their corporate mission in response to an ecosystem increasingly complex, a broader and non-homogeneous stakeholders base with different objectives, investment horizons and that assesses performance and value based on different methods?

Does the Conscious Corporation Need a Dedicated Legal Framework to Thrive? Initial Observations

While speaking about these interesting issues we need to fix and summarize the following main concepts: (a) any corporation that wishes to be part of the Climate Change Ecosystem should be a self-conscious corporation. It should be aware of its role within the ecosystem and of the consequences that its actions will inevitably have on the ecosystem itself. The conscious corporation is an integral part of its ecosystem and due to the intimate relationship between the various component of the ecosystem, decoupling is not possible.

As eloquently indicated (Pope Francis, 2015) "*it cannot be emphasized enough how everything is interconnected. Time and space are not independent of one another, and not even atoms or subatomic particles can be considered in isolation. Just as the different aspects of the planet – physical, chemical and biological – are interrelated, so too living species are part of a network which we will never fully explore and understand*".

As a consequence, a corporation will never be able to thrive in the long term if it failed to create value for the ecosystem as a whole; (b) the conscious corporation should have a long-term view. Short termism is not only negative but more fundamentally inconsistent with the rationale underpinning sustainable investing. A corporation that generates short term profits without assessing the long-term sustainability of its business is most likely compromising its ability to generate value in the long term; (c) transparency and homogenous data disclosure are essential pre-requisite for the conscious corporation to be able to attract new capital, and (d) the governance model should protect and advance the objectives of all the stakeholders in the ecosystem and enable the ecosystem as whole to thrive.

Would a dedicated legal framework be necessary or beneficial for the conscious corporation? Does the absence of a dedicated legal framework negatively affect the ability of the conscious corporation to achieve its objectives and pursue its corporate mission? The answer to the question is largely dependent on the characteristics of the legal framework in question. However, we will only limit our preliminary and not exhaustive analysis to the theoretical benefits of a dedicated legal framework irrespective of its specific characteristics.

In the US and Europe lawmakers have made several attempts to regulate social enterprises and benefit corporations. Today at least nineteen states in North America have passed a legislation on social enterprises.

On the other side of the Atlantic, the UK and Italy have recognized the concept of Community Interest Companies and "*società benefit*", respectively. In 2009 the concept of mission corporation (Société a' mission) was introduced in France by the Loi Pacte.

A Community Interest Company is a limited liability company designed for social enterprises which has the specific aim of providing benefit to a community and uses its income, assets and profits for the community it is formed to serve. It can be limited by shares or by guarantee but must satisfy a community interest test. This could potentially be a useful model for a future hypothetical dedicated legal framework assuming, of course, that the interest satisfied by the corporation is wide enough to be relevant for the community as a whole and not just for a limited group of stakeholders and the corporation satisfies the "do no harm test". Such a test requires that the activity of the corporation should not harm any of the other objectives of the community.

The Italian "*società benefit*" (SB) was introduced in Italy by Law 208/2015. The Italian SB is a for-profit corporation with a dual corporate object. It conducts a traditional economic activity but simultaneously attempts to pursue a common benefit. The Italian model is a first in a civil law environment.

According to the currently prevailing views on benefit corporations (Battilana et al., 2017) Italian SBs, due to their dual corporate object, may be classified as hybrid organizations. Pursuing simultaneously a social purpose or an environmental objective and profitability is a challenging exercise. The different objectives to be pursued by a benefit corporation, or a società benefit maybe conflicting due to their very nature or simply because, at least in the short term, they are inconsistent with the for-profit status of the corporation. Simply put, the for-profit status may conflict with the pursue of a social purpose. In order to manage such conflicts, a governance enabling a transparent and predictable allocation of priorities is essential. New skills may have to be developed in order to fully exploit the potential of the SB status. The law for example provides for a benefit officer to be appointed in order to ensure, among other things, the harmonized implementation of the objectives of the corporation.

A recent study on Italian benefit companies (Bellavite Pellegrini & Caruso, 2020) produced data indicating that the number of newly formed società benefit and of existing corporates electing to convert to the SB status has been increasing since the implementation of the SB framework in Italy. More importantly, the study produced empirical evidence demonstrating that the SB status "has a positive effect on ROA" and "may contribute to a reduction of risk and of cost of capital".

It may be argued that the dual corporate object of Italian SBs makes them a good fit for the Climate Change Ecosystem. It is also inherently consistent with the theory according to which it is possible for firms to do well while doing good. Empirical evidence has been collected by scholars Fan and Michalski (2020) showing that sustainable investing allows incorporating ethical preferences while offering strong potential for wealth generation. As a result, the SB model appears to be in principle the right tool to attract capital from investors—whether or not such investors have a mandate to invest in sustainable or ESG projects and assets—and direct such capital to sustainable investments and projects. Directing financial resources to sustainable projects and investments is a priority for European environmental policies (see Chapter 2). Due to the above, it may be legitimate to assume that a model that is financially sound and an enabler of policy objectives is likely to be replicated and fine-tuned.

The benefit of a legal framework, like any other working tool, should not be assessed in principle but by reference to its ability to contribute to the achievement of a pre-determined result.

A conscious corporation creates value by constantly increasing its awareness of the objectives and priorities of the Climate Change Ecosystem, by fostering knowledge, research and innovation, by deploying financial capital towards projects that would enable the ecosystem to achieve its objectives. The Climate Change Ecosystem is dynamic and is constantly evolving.

Requiring a corporation, like the conscious corporation, whose major strengths include consciousness and awareness and the ability to use discretion for the benefit of all stakeholders to act based on a legal framework is not without challenges. The ability to adapt and exercise discretion are fundamental conditions for the conscious company to create value. Any restriction imposed on the ability of the conscious corporation to adapt and exercise discretion would very likely jeopardize the ability of the corporation to create long term value. Whether or not a dedicated legal framework could facilitate or encourage the adaptation or evolution of the conscious corporation is yet to be determined and further research would be necessary in order to express a final view on this matter. In this respect, it has been observed (Bellavite Pellegrini & Caruso, 2020) that the Italian SB framework contributes to clarify the relation between the different components of the corporate object of SBs and therefore should be considered as a valuable tool for transitioning to a more sustainable economy.

It is however evident that the knowledge of ESG matters enables a thorough understanding of the risks and opportunities a company faces, allowing enhanced security selection and risk management. Additionally, ESG analysis leads to an improved understanding of how future trends could affect a certain industry or the entire economic landscape for that matter. The second part of the book will explore and provide empirical evidence supporting the theory whereby corporations that have invested in adaptation and with a higher ESG score typically create more value than their peers that have failed to adapt or are less advanced in their adaptation process.

References

Amel-Zadeh, A., & Serafeim, G. (2018). Why and how investors use ESG information: Evidence from a global survey. *Financial Analysts Journal, 74*(3), 87–103.

Battilana, J., Besharov, M., & Mitzinneck, B. (2017). On hybrids and hybrid organizing: A review and roadmap for future research. *The SAGE Handbook of Organizational Institutionalism, 2*, 133–169.

Behrens, A. (2016). *Time to connect the dots: what is the link between climate change policy and the circular economy.*

Bellavite Pellegrini, C., & Caruso, R. (2020). *Societa' Benefit, Profili giuridici ed economico-aziendali.* Egea.

Berger, A. N., Imbierowicz, B., & Rauch, C. (2016). The roles of corporate governance in bank failures during the recent financial crisis. *Journal of Money, Credit and Banking, 48*(4), 729–770.

Bodansky, D., Brunnée, J., & Rajamani, L. (2017, June 10). *Introduction to international climate change law.* International Climate Change Law (Oxford University Press), Available at SSRN: https://ssrn.com/abstract=3000009 or http://dx.doi.org/10.2139/ssrn.3000009

Burris, S., Drahos, P., & Shearing, C. (2005). Nodal governance. *Australian Journal of Legal Philosophy, 30*, 30.

Caruso, R., Petrarca, I., & Ricciuti, R. (2016). Climate change, rice crops and violence: Evidence from Indonesia. *Journal of Peace Research, 53*(1), 66–83.

Darwin, C. (1859). *On the origin of species by means of natural selection or the preservation of favoured races in the struggle for life.* International Book Company.

Fan, J. H., & Michalski, L. (2020). Sustainable factor investing: Where doing well meets doing good. *International Review of Economics & Finance, 70*, 230–256.

Francis, P. (2015, May 24). *Laudato si.* Vatican Press, w2.

Frentrop, P. (2012). *Short-termism of institutional investors and the double agency problem*. Available at SSRN: https://ssrn.com/abstract=2249872 or http://dx.doi.org/10.2139/ssrn.2249872

Friedman, M. (1962). *Capitalism and freedom* (p. 133). The University of Chicago Press.

Guay, W. R., Taylor, D. J., & Xiao, J. J. (2015). *Adapt or perish: Evidence of CEO adaptability to industry shocks*. Available at SSRN 2234886.

Harper, K. (2017). *The fate of Rome*. Princeton University Press.

Kwon, S. W., & Arenius, P. (2010). Nations of entrepreneurs: A social capital perspective. *Journal of Business Venturing, 25*(3), 315–330.

Lashitew, A. A., & Werker, E. (2020). Do natural resources help or hinder development? Resource abundance, dependence, and the role of institutions. *Resource and Energy Economics, 61*, 101183.

Maniruzzaman, A. F. M., & Al-Saleem, K. (2017) The energy and environment dilemma: Sustainably developing Iraqi oil and gas in International law and policy-prospects and challenges. *Oil, Gas & Energy Law Intelligence*.

Mayer, C. (2018). *Prosperity: Better business makes the greater good*. Oxford University Press.

Meadowcroft, J. (2009). What about the politics? Sustainable development, transition management and long-term energy. *Transitions Policy Sciences, 42*(4), 323–340.

OECD. (2011). *Benefits of investing in water and sanitation*.

OECD. (2020). *Evaluating the initial impact of COVID-19 containment measures on economic activity*.

Patel, A., Singh, P. K., & Tandon, S. (2010). *Weather prediction using machine learning* (February 2021). Available at SSRN: https://ssrn.com/abstract=3836085 or http://dx.doi.org/10.2139/ssrn.3836085

San-Akca, B., Sever, S. D., & Yilmaz, S. (2020). Does natural gas fuel civil war? Rethinking energy security, international relations, and fossil-fuel conflict. *Energy Research & Social Science, 70*, 101690.

Savaresi, A. (2016). The Paris Agreement: A new beginning? *Journal of Energy & Natural Resources Law, 34*(1), 16–26. https://doi.org/10.1080/02646811.2016.1133983

Schumpeter, J. A. (1976/1942). *II. Capitalism, socialism, and democracy*. Routledge.

Smith, B. (1928). Legal personality. *The Yale Law Journal, 37*(3), 283–299.

World Commission. (1987). *Report of the on environmental development: Our common future*.

CHAPTER 3

Climate Change Challenges and the Policymakers Initial Response

Massimo Catizone

Introduction: A "Green" Field of Study

Interest in nature and the environment is certainly not a novelty. In the eighteenth century, romantic poets cultivated a reverence for nature as it was regarded as a pure spiritual source of renewal. When hiking in the Lake District in England, the poet William Wordsworth wrote that nature is a "*sort of national property in which every man has a right and interest who has an eye to perceive and a heart to enjoy*".

It may be argued that Wordsworth's view remains relevant to our days. However, we have now moved beyond the romantic vision of nature. The way in which we look at our planet, its resources, the relationship between environmental, social and governance factors is more complex, nuanced and often conflictual, as there is no unanimous consensus on the magnitude of climate change-related risks and on how they should

M. Catizone (✉)
Research Centre of Applied Economics, Catholic University of the Sacred Heart, Milan, Italy

© The Author(s), under exclusive license to Springer Nature Switzerland AG 2022

C. B. Pellegrini et al. (eds.), *Climate Change Adaptation, Governance and New Issues of Value*, Palgrave Studies in Impact Finance,
https://doi.org/10.1007/978-3-030-90115-8_3

be mitigated and managed. Environmental social and governance factors have only very recently and to a limited extent started to feature in corporate strategies, investment strategies and financial regulation. The majority of corporations and investment managers do not fully appreciate the impact of climate change on their business and do not incorporate, or do so only marginally, in their business model and investment strategies, sustainability considerations.

This poses new challenges and the attempts made by the scientific community, the industry, lawmakers and regulators to reach a consensus on how these challenges should be addressed has led to the emergence of a new field of study focussed on sustainability, climate change resilience, environmental policy integration and the impact of environmental, social and governance factors on the global economy and financial stability.

The purpose of this chapter is to provide an overview of the main principles underpinning the international regulatory framework applicable to climate change, how and why such principles have been enshrined in European regulations and how European lawmakers and regulators intend to support and serve the Climate Change Ecosystem.

From High-Level Policy Objectives to Investible Assets

The review of the existing international regulatory framework and of the set of principles intended to discipline the Climate Change Ecosystem can be based on different approaches.

The starting point of our proposed approach is that the transition to a greener, circular, more resilient and sustainable economy can only be achieved if sufficient capital is invested to support such transition. In this chapter we are proposing a review of the main components of the current regulatory framework based on their aptness to convert policy objectives into investible assets.

The financial resources that can be made available by the public sector to facilitate the transition to a greener economy are not sufficient for such transition to take place. Therefore, private sector stakeholders, including financial product manufacturers, investors and distributors have a very important role to play as they can reorient capital towards sustainable investments, virtuous corporations, new technologies and projects contributing to the creation of a low-carbon, climate-resilient and circular economy.

A Brief History

The establishment of The Intergovernmental Panel on Climate Change (IPCC) in 1988 can be regarded as an early signal that a form of consensus around the need to understand and tackle the challenges posed by climate change was gradually forming. IPCC was created by the World Meteorological Organization and the UN Environmental Programme to provide a scientific assessment of climate change. The role of IPCC is largely unchanged and IPCC remains to this day a key driver behind climate change-related policymaking. It has been argued (Boehmer-Christiansen, 1996) that the close link between science and politics that characterizes the IPCC could potentially negatively affect the scientific value of its positions and assessments. The first assessment report issued by the IPCC highlighted that the emissions resulting from human activities were increasing atmospheric concentrations of greenhouse gases and that a global treaty was necessary to address the issue. In December 1990 the Intergovernmental Negotiating Committee (INC) was created in order to agree a framework convention on climate change. In May 1992, after five sessions during which delegates from 150 states discussed emission reductions, technology transfer, roadmaps and targets for emissions reduction, the United Nations Framework Convention on Climate Change (UNFCCC) was adopted. It entered into force in March 1994.

UNFCCC acknowledged that "change in the Earth's climate and its adverse effects are a common concern for humankind" and introduced, for the first time, concepts and mechanisms that to this day are shaping climate-related policies, laws and regulations. Article 2 of UNFCCC clearly states that "stabilization of greenhouse gas concentrations in the atmosphere at a level that would prevent dangerous anthropogenic interference with the climate system" is the ultimate objective of UNFCCC. The article goes on stating that such level should be implemented at a pace which would allow ecosystems to adapt without threatening or disrupting food production and economic activities. It can be argued that this is the intuition underpinning the concept of transition risk. Today the relevance of transition risk is generally acknowledged and the stakeholders of the Climate Change Ecosystem are devoting significant resources in attempting to predict, quantify and mitigate transition risk. However, at the time of the UNFCCC the reference to transition risk was innovative and it certainly contributed to focus scholars' and regulators' attention on the components and possible effects of such risk on the global economy. For the first time, again, Article 4 introduced the concept that the parties to the UNFCCC have common but different responsibilities, hence suggesting that climate change is a global challenge, but its impact and the ability to prevent and manage it vary significantly from country to country. Furthermore, Article 4 (e) requires the parties to cooperate in preparing for adaptation. Adaptation, as we have discussed in Chapter 1 is a key component of the conscious corporation and it is essential for the Climate Change Ecosystem to function. As we will discuss below, adaptation is also prioritized and encouraged by the current European regulatory framework on climate change.

UNFCCC established the Financial Mechanism (Article 11) and the Conferences of Parties (COP) (Article 7).

The Financial Mechanism is intended to serve the Kyoto Protocol and the Paris Agreement and provides climate finance consistently with the principle, set out in the convention, according to which parties have common, but differentiated responsibilities. The term climate finance is not very well defined; however, it is typically used in the UN documentation to describe national and transnational financing—drawn from public, private and alternative sources of financing—that seek to support mitigation and adaptation actions that will address climate change.

The Global Environment Facility was the first operating entity to be established under the Financial Mechanism. In 2010, during COP 16, the Green Climate Fund was established. Today the parties to the convention have also established the Special Climate Change Fund, the Least Developed Countries Fund and the Adaptation Fund.

The COPs are the supreme body of the convention and are the formal meetings of the parties to the UNFCCC. In essence, COPs have two macro priorities: advance the objective of the UNFCCC and assess the progress made by the parties in dealing with climate change. COPs commenced their activities in the 1990s with the negotiation of the Kyoto Protocol intended to establish legally binding obligations for developed countries to reduce their greenhouse gas emissions. Since then the COP met each year. Particularly noteworthy was the outcome of COP 21 also known as the 21st Conference of Parties which was held in Paris from 30th November to 12th December 2015. Negotiations resulted in the Paris Agreement. On 4th October 2016 the threshold for adoption was reached with over 55 countries representing 55% of the world's greenhouse emission ratifying the agreement. Arguably, the adoption of the Paris Agreement, also known as the Paris Accord on Climate Change, marks the beginning of the contemporary history of the fight against climate change.

The Paris Agreement includes the commitment to align financial flows with a pathway towards low carbon and climate-resilient development. In this respect the Nationally Determined Contributions (NDCs) can be regarded as one of the cornerstones of the agreement. Pursuant to Article 4, paragraph 2, each party is expected to determine, communicate and maintain nationally determined contributions. These are objectives that each party voluntarily commits to meet in order to contribute to the achievement of the overall objectives of the Paris Agreement. Each country NDCs are therefore a reflection of such country level of commitment and actual contribution towards emissions reduction.

The objectives set out in the agreement are ambitious and over the years their implementation has not always been free from controversy. In 2017, for example, President Trump announced the intention of the United States to withdraw from the Paris Agreement.

The United Nations (UN) 2030 Agenda for Sustainable Development (Agenda) setting out the Sustainable Development Goals (SDGs) was also adopted in 2015. The Agenda is the result of years of international joint efforts made by the United Nations and several countries. It is widely regarded as a major stepping stone to a global integrated sustainable economy. Mathers and Deonandan (2018) expressed a contrarian view and argued that the SDGs should put more emphasis on governance and other tasks aimed at curtailing corruption and tax evasion. More than five years after its adoption, it seems legitimate to argue that the SDGs set out in the Agenda have driven international cooperation and helped countries and intergovernmental bodies across the world shaping their climate change policies.

In the following paragraphs of this chapter we will attempt to illustrate how the Agenda has influenced European policymakers and regulators (Fig. 3.1).

The SDGs are very ambitious objectives. Due to their complexity it is evident that they can only be achieved by virtue of a dynamic process involving constant capital injections, technical innovation and an effective international cooperation mechanism aimed at enhancing climate change capabilities, encouraging knowledge sharing and ultimately leading to the harmonization of climate change-related policies. A single event or action is not going to be sufficient to achieve this objective. An ongoing and carefully managed process is required. The

Fig. 3.1 UN Sustainable Development Goals (*Source* UN Sustainable Development Goals)

UN appears to be fully committed to ensure the success of such process. The Division for Sustainable Development Goals in the United Nations Department of Economic and Social Affairs is actively involved in facilitating the implementation of the SDGs by providing analytical support aimed at facilitating capacity development and stakeholders' engagement and outreach.

It can be argued that the Agenda is relevant for several reasons that go well beyond its 17 Sustainable Development Goals and 169 targets. First, it acknowledges that the goals are integrated and indivisible. Sustainable development can only be achieved in a nonviolent environment where human rights are protected. Equally, an environment where human rights are not adequately protected is likely to be affected by conflicts, migrations and other social and financially destabilizing events.

Second, the launch of Technology Facilitation Mechanism (TFM) intended to facilitate multi-stakeholder collaboration and partnership through the sharing of information, scientific knowledge and best practices. The TFM rests on three pillars: (a) a United Nations Interagency Task Team on Science, Technology and innovation for SDGs; (b) a collaborative Multi-stakeholder Forum on Science, Technology and Innovation for SDGs (STI Forum) and (c) an online platform as a gateway for information on existing STI initiatives, mechanisms and programmes.

Last but not least, the Agenda contains a formal acknowledgement of the fundamental contribution that may be provided by the private sector to the advancement of the SDGs both in terms of resources mobilization and as a source of valuable creativity, knowledge and innovation.

These are key principles that will reappear in other international policies designed to address the challenges posed by climate change. The European Union is no exception.

In 2016, the European Commission released a communication discussing its rationale and strategy for implementing in the European legislation the principles contained in the SDGs. In the communication the Commission recognizes that the Agenda is fully consistent with Europe's vision and that all countries, with no exception are responsible for achieving the SDGs. Furthermore, in the communication, the European Commission indicates that the Agenda reflects for the first time an international consensus that peace, security, justice for all and social inclusion are not only to be pursued on their own but that they reinforce each other. This is very much consistent with the acknowledgement contained in the Agenda that the SDGs are integrated and indivisible.

A mapping of each SDGs against then-existing European policies was also contained in the communication. The outcome of this exercise—that is set out in the Staff Working Document accompanying the document—was twofold. First, it became apparent that most of the goals were already in the process of being implemented. The political agenda of the Commission at the time (Junker Commission) was driven by the principles set out in the Strategic Agenda for the Union in Times of Change and accordingly employment, growth, fairness and democratic change featured quite predominantly. Also, it should not be surprising that the SDGs were regarded as being fully consistent with the ten priorities set out in the agenda. Second, it was acknowledged that in order to ensure a full and harmonized implementation of the SDGs, further actions were required. Table 3.1 summarizes the correspondence between the priorities set out in the Junker Strategic Agenda and the SDGs.

As a further response to the Agenda, the European Union launched a reflection work on how to implement the SDGs in the long term. This led to the publication of a reflection paper intended, among other things, to contribute to the European Union's Strategic Agenda 2019–2024.

Table 3.1 Mapping Junker Commission Strategic Agenda against the Sustainable Development Goals

Junker Commission Strategic Agenda	*Sustainable Development Goals*
Priority 1—New boost for jobs, growth and investment	SDG4: Quality Education SDG 8: Decent Work and Economic Growth SDG 10: Reduced Inequalities
Priority 2—A Connected Digital Single market	SDG 9: Industry Innovation and Infrastructure SDG 12: Sustainable Cities and Communities
Priority 3—A resilient energy union with a forward-looking climate change policy	SDG 7: Affordable and Clean Energy SDG 13: Climate Action SDG14: Life Below Water SDG 15: Life on Land
Priority 4—A deeper and fairer international market with a strengthened industrial base	SDG 12: Responsible Consumption and Production
Priority 5—A deeper and fairer Monetary Union	SDG3: Good Health and Well-Being SDG 1: No Poverty SDG2: Zero Hunger SD6: Clean Water and Sanitation
Priority 6—A Reasonable and balanced free trade agreement with the US	–
Priority 7—An area of justice and fundamental rights based on mutual trust	SDG 16: Peace Justice and Strong Institutions SDG 5: Gender Equality SDG 17: Partnerships for the Goals
Priority 8—Towards a new policy on migration	SDG 8: Decent Work and Economic Growth SDG 10: Reduced Inequalities SDG3: Good Health and Well-Being SDG 1: No Poverty SDG2: Zero Hunger SD6: Clean Water and Sanitation
Priority 9—A stronger global actor	SDG 17: Partnership for the Goals
Priority 10—A union of democratic change	SDG 16: Peace Justice and Strong Institutions

Source Author's elaboration

The paper is a relatively long document (132 pages) and proposes three different courses of action which in the paper are described as scenarios:

- Scenario 1: an overarching EU SDG strategy to guide all actions of the EU and its member states;

- Scenario 2: Continued mainstreaming of the SDGs in all relevant EU policies by the Commission, but enforcing Member States action; and
- Scenario 3: Putting enhanced focus on external action while consolidating current sustainability ambition at EU level.

Our review of the subsequent initiatives taken at the European level seems to suggest that, whether deliberately or not, Europe is favouring an implementation strategy that combines the three scenarios.

In December 2016 the Commission established the High-Level Expert group on sustainable finance (HLEG). The HLEG, comprising of 20 members from academia and experts from different fields was mandated to develop an EU strategy on sustainable finance and advise the European Commission on how to channel financial resources towards sustainable investments; identify the steps that financial institutions and supervisors should take to protect stability of the financial system from risks related to the environment, and ensure the harmonized application of any policy intended to implement the above-mentioned objectives.

One year later, on January 2018, HLEG published its final report. The recommendations set out in the HLEG final report can be summarized as follows:

1. to introduce a common sustainable finance taxonomy to ensure market consistency and clarity, starting with climate change;
2. to clarify investors duties to extend time horizons and bring greater focus on ESG factors;
3. to upgrade Europe's disclosure rules to make climate change risks and opportunities fully transparent;
4. to empower and connect Europe's citizens with sustainable finance issues;
5. to develop official European sustainable finance standards, starting with one on green bonds;
6. to establish a "Sustainable Infrastructure Europe" facility to expand the size and quality of the EU pipeline of sustainable assets;
7. to reform governance and leadership of companies to build sustainable finance;
8. to enlarge the role and capabilities of the ESAs to promote sustainable finance as part of their mandate.

The above recommendations can be divided into two main macro-areas and objectives: (a) improve the contribution of finance to sustainable and inclusive growth and (b) strengthen financial stability by incorporating environmental, social and governance factors into investment decision making. Below we will discuss how the recommendations of the HLEG, continue to shape European policymaking on climate change.

To follow up on the works of the HLEG the Commission published an Action Plan on Financing Sustainable Growth. The Plan, which may be regarded as a further effort to use finance as a tool to advance the macro long-term objectives of the European economy sets out three main objectives: (a) reorient capital flows towards sustainable investment in order to achieve sustainable and inclusive growth; (b) manage financial risks stemming from climate change, resources depletion, environmental degradation and social issues and (c) foster transparency and long-termism in financial and economic activity.

These objectives are to be achieved by implementing ten main actions. Considering each action in turn:

Action 1: Establishing an EU classification system for sustainable finance. The Commission acknowledged that the lack of a harmonized classification setting out which activities can be regarded as sustainable and what actions can contribute to climate change mitigation or adaptation is one of the main factors negatively affecting the flow of capital, including private capital, towards sustainable activities. The Technical Expert Group, comprising 35 members from civil society, academia, business and the finance sector as well as additional members and observers from EU and international public bodies, was established by the Commission in order to assist with the development of a unified classification of sustainable economic activities, an EU green bond standard, and methodologies for low-carbon indices. It provided its contribution to the solution of the issues identified by the Commission, in The Taxonomy Technical Report released in June 2019. This report, commonly referred to as the EU Taxonomy, is a more than 400 pages document divided into six parts setting out (a) the taxonomy approach; (b) the methodology used for developing technical screening criteria for climate change mitigation objectives, (c) adaptation objectives and as we have indicated in Chapter 1, it also details the rationale for the "do no significant harm" principle; (d) practical guidance to potential users; (e) an initial assessment of the potential impact of establishing an EU Taxonomy; (f) next steps and (g) a full list of technical screening criteria. In April

2020 the European Council adopted a regulation setting out an EU classification system intended to identify those activities which are considered environmentally sustainable. The future framework will facilitate the advancement of the following objectives: (a) climate change mitigation, (b) climate change adaptation, (c) sustainable use and protection of water and marine resources, (d) transition to a circular economy and (e) pollution prevention and control, protection and restoration of biodiversity and ecosystems. The application of the six principles should be fully phased in by 2022. The EU Taxonomy and its technical standards are expected to become fully operational in January 2022.

Action 2 Creating standards and labels for green financial products. In this context the Technical Expert Group was asked *to advise on a green bond standard and explore the merits of using the EU Ecolabel framework to advance the development of green financial products.*

Figure 3.2 shows that the majority of sustainable investments were made in the public equity sector.

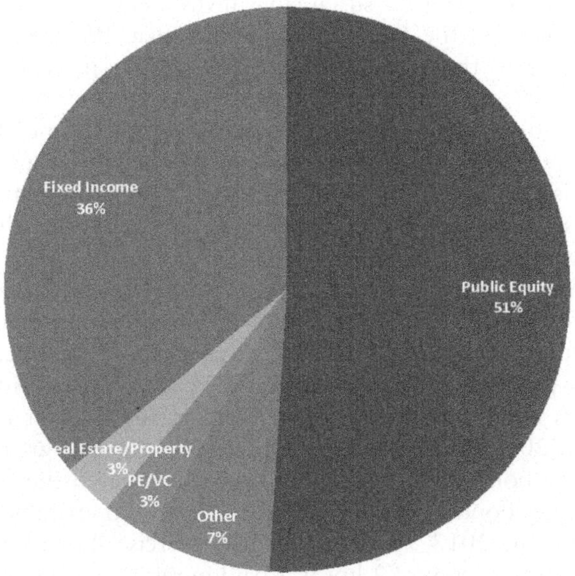

Fig. 3.2 Global sustainable investing (*Source* Global Sustainable Investment Alliance)

It may be argued that the propensity to invest in sustainable fixed income products has been negatively affected by several factors, including:

1. Uncertainty surrounding the concept of green bond and sustainable investments. Historically, the uncertainty surrounding the concept of sustainable investment has negatively affected the growth path of the sector. As we have mentioned before, the recent European regulation on taxonomy defines what constitutes a sustainable investment. This is expected to be beneficial as it should help in expanding the range of green eligible investments and reduce the potential for whitewashing and social washing, hence ultimately boosting investors' confidence in the sector.
2. Policymakers attitude towards sustainable investments. The position of policymakers around the world varies significantly. In Europe, one of the primary objectives of climate policies is directing new capital towards sustainable investments and large public investors like pension funds are required to allocate a portion of their assets under management to sustainable investments. The stance of the policymakers in the UK is not too dissimilar. We are not aware of a provision to the same effect in the US where it has recently been made clear to pension funds that they are under no obligation to allocate to sustainable investments. As a result, in the absence of a radical policy change, the decision to allocate to sustainable investments is expected to be solely driven by the investment mandate of each asset manager and the customary assessment of the risks and the rewards.
3. Greenium. This is the term often used by market participants to indicate the green premium payable in respect of a green bond. The green premium may be positive or negative. Should green bonds carry a positive or negative premium relative to traditional bonds? Scholars have studied the subject. Ehlers and Parcker (2017), and Zerbib (2019) have observed an average negative green premium for green bonds. More precisely Zerbib compared the performance of a green bond and the performance of a comparable traditional bond for the 2013–2017 period. The exercise showed that a negative green premium of 2 bps was paid in respect of the green bonds. Fatica et al. (2019) found that a negative premium was paid in respect of bonds issued by supranational and corporate issuers and that no negative premium was paid in respect of bonds issued by

financial institutions. Baker et al. (2018) showed that US green municipal bonds typically paid a negative premium of 7 bps relative to comparable traditional bonds. Secondary market observations have also yielded conflicting results. Kerpf and Mandel (2018) have observed that the yield curve of US municipal green bonds is tighter than the yield curve of traditional municipal bonds. On the contrary Hachenberg and Schiereck (2018) reviewed a basket of investment-grade bonds and were able to observe a small positive premium. Ghoul et al. (2011) demonstrated that high ESG ratings, typically have the benefit of facilitating access to finance, but non-conclusive evidence emerged in respect of pricing.

4. Liquidity and cannibalization risk. Part of the issuers' community is concerned that by building a green curve an issuer may inadvertently reduce the liquidity and depth of its traditional bonds. In order to mitigate this risk, in its September 2020 issuance, the German Federal Ministry of Finance, devised a solution that became known in the market as the Green Twin. The conventional bonds and their green twin have the same coupon and maturity (Fig. 3.3).

In June 2019, the TEG produced a report setting out a proposal for an EU Green Bond Standard. The report impinges on ten recommendations, three of which relate to the establishment of the EU green bond standards and the others relate to ways in which stakeholders may support the harmonized implementation of the European Green Bond Standards, namely: (1) create a voluntary green bond standard; (2) the EU Green Bond Standards should comprise four core components: (a) alignment of green projects with the EU Taxonomy; (b) Green Bond Framework, (c) reporting and (d) verification by accredited verifiers; (3) encourage the set up of a voluntary interim registration process for verifiers of EU Green Bonds for an estimated transition period of up to three years; (4) encourage investors to use EU-GBS when designing green fixed income strategies and to communicate their preference and expectations actively to green bond issuers and distributors; (5) adoption of a disclosure regime on green bonds holding for institutional investors; (6) promoting greening the financial system by expressing and implementing a preference for EU Green Bonds; (7) developing financial incentives to support the EU Green Bonds Market alignment with EU-GBS; (8) encouraging issuers to issue their green bonds in conformity with the requirements of the EU-GBS; (9) promote adoption of the EU-GBS through the EU

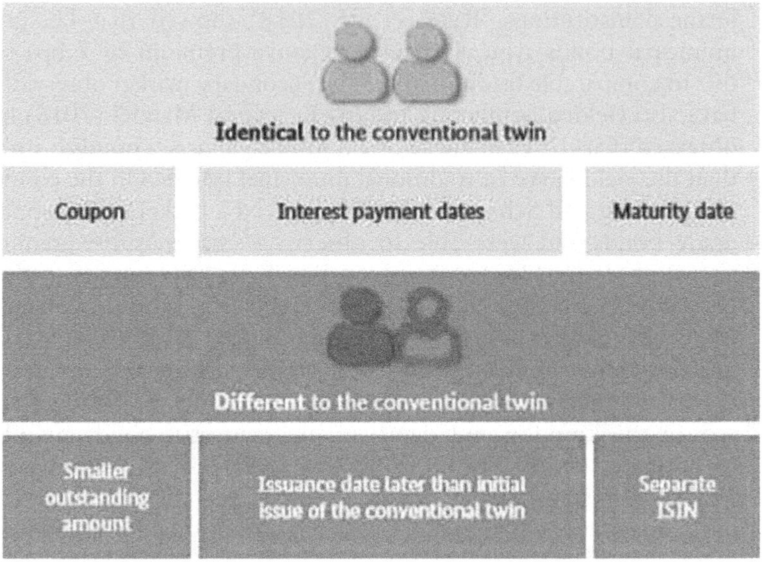

Fig. 3.3 Germany green twin bonds (*Source* Green Bond Framework, Federal Ministry of Finance [Germany] and NN Investment Partners)

Ecolabel for financial products; (10) monitor the impact on the alignment of financial flows with the EU Taxonomy's Environmental Objectives and consider further supporting actions including possible legislation after an estimated period of up to three years.

The private sector is also actively contributing to the advancement of the EU Green Bond Standards recommendations with specific initiatives aimed at developing climate finance solutions in the real estate market. For example, the Sustainable Covered Bond Label devised by the European Covered Bond Council is a quality label aimed at increasing transparency in the covered bonds market. The label is only assigned following a specific authorization process and provided that the issuer commits to use "an amount equivalent to the proceeds of that same covered bond to (re)finance loans in clearly defined environmental and or social" projects. There are additional projects with a similar objective. The position paper published by the Association for Financial Markets in Europe, Principles for developing a green securitization market in Europe (2019) setting out the principles that AFME believe regulators should support to encourage green securitization is another indicator that stakeholders have started to fully appreciate the principles and the benefits of the Green Bonds Principles. The Green Bond Principles, Voluntary Process Guidelines for Issuing Green Bonds (2018) published by the International Capital Markets Association (ICMA) is another important step in the same direction.

Action 3 Fostering investment in sustainable finance; and Action 4 Incorporating sustainability when providing financial advice. Advice provided by investment firms is deemed to be crucial in order to redirect capital towards sustainable projects. The Markets in Financial Instruments Directive (MiFID II) and the Insurance Distribution Directive require investment firms and distributors to take into account clients appetite, risk tolerance and investment objectives in order to identify the most suitable products. It is likely that amendments to MiFID II and IDD will be introduced in order to ensure that perspective investors' preferences in respect of sustainability and environmental social and governance factors are duly considered.

Action 5 Developing Sustainability Benchmarks. This action is intended to create more transparent methodologies for developing indices. The purpose of these indices is twofold. First, channelling capital towards sustainable investments. Second, reducing greenwashing risk. In April 2020 the benchmark regulation was amended. The Commission introduced two new climate-related benchmarks: The EU Climate Transition Benchmark and the EU Paris-aligned benchmark. The new benchmarks are expected to provide investors with better information on the carbon footprint of their investments.

Action 6 Better integrating sustainability in ratings and market research. An improved integration of environmental social and governance factors in the rating agencies assessment of credit risk is perceived to be conducive to an asset allocation weighted towards sustainable projects. In this respect the Commission proposed to (a) assess the need for amending the Credit Rating Agencies Regulation; (b) conduct a comprehensive study of sustainability ratings and of the underlying methodologies and (c) invite ESMA to include environmental and social sustainability information in its guidelines on disclosure for credit rating agencies.

Action 7 Clarifying institutional investors' and asset managers duties. This action is aimed at addressing the concern that sustainability factors and risks were not taken duly into account in investment decisions. It may be argued that clarifying institutional investors and asset managers duties in respect of sustainability considerations may, among other things, facilitate compliance with fiduciary duties.

Action 8 Incorporating sustainability in prudential requirements. In this respect the Commission elected to explore the feasibility of the inclusion of risks associated with climate and other environmental factors in institutions' risk management policies and the potential calibration of capital requirements of banks as part of the Capital Requirements Directive. Furthermore, the Commission is also expecting the European Insurance and Occupational Pensions Authority (EIOPA) to provide an opinion on the impact of prudential rules for insurance companies on sustainable investment ultimately with the view of assessing the need to amend Solvency II Directive.

Action 9 Strengthening sustainability disclosure and accounting rule-making. This action is intended to enable investors and other stakeholders assessing companies' long-term value creation. Disclosure of non-financial risks was not new at the time, but the ultimate objective of this action was assessing the relevant accounting rule with the view of striking a balance between the investors need for data comparability and corporates ability to exercise discretion when disclosing risks. Furthermore, this action is also aimed at assessing whether the existing accounting rules contain any provision directly or indirectly causing corporations to prioritize short-term gains at the expenses of long-term value creation. In order to support the Commission with the implementation of this action, in 2019 the TEG published the Report on Climate-Related Disclosures. The report contains guidance intended to assist corporates in developing high-quality climate-related disclosures that comply with the requirements of Directive 2014/95, also known as the Non-Financial Reporting Directive (NFRD) which is the legal basis for the implementation of the Task Force on Climate-related Financial Disclosure (TCFD) recommendations on this subject. The TCFD was established by the G20 Financial Stability Board. TCFD is intended to develop voluntary, consistent climate-related financial risk disclosures requirements to be met by investors, lenders, insurers and other stakeholders. It is hoped that the work of TCFD, which represents corporations with assets shy of USD 9TN, will help corporations in gauging market expectations and adapt their disclosure standards accordingly. A proposal for a regulation on disclosure relating to sustainable investments and sustainability risks and amending Directive

(EU)2016/2341 has also been made in an attempt to advance this action. The proposed regulation, if implemented, should introduce disclosure obligations on how institutional investors and asset managers integrate environmental, social and governance factors in their risk management processes. At present a new green asset ratio (GAR) intended to measure the ratio between total assets and Taxonomy aligned assets is under discussion. If implemented, it may prove a valuable transparency enabler for financial institutions.

Action 10 Fostering sustainable corporate governance and attenuating short-termism in capital markets. History shows that corporate governance is intimately connected with sustainability and long-term value creation. This action was intended to encourage the adoption of governance mechanisms conducive to sustainable investments. It is generally recognized that firms are aware of the relation between corporate performance and sustainability and high-quality disclosure. However, there appears to be limited integration of sustainability concepts into corporate governance mechanisms (Krechovská & Procházková, 2014). Furthermore, as the Climate Change Ecosystem is dynamic, corporate governance should also evolve. It has also been argued (Levillain & Segrestin, 2019) that in some cases it may be difficult for traditional corporate bodies to fully grasp the specificities of environmental, social and governance factors and therefore corporate governance and the pursuit of the corporate object should be tasked to different internal bodies and functions. The Commission committed to conduct an assessment to establish the need for corporate boards to develop and disclose a sustainable strategy and the possible need to clarify the rules according to which directors are expected to act in the company's long-term interest. There is a consensus that board composition affects financial performance. However, recent empirical evidence (Naciti, 2019) based on data relating to board diversity, board independence and CEO duality shows that firms with more diversity on the board and a separation between chair and CEO roles show higher sustainability performance. A higher number of independent directors leads to lower sustainability performance.

Figure 3.4 illustrates the connection between each action and the main objectives of the Action Plan.

In October 2019 the EU launched the International Platform on Sustainable Finance (IPSS) in the presence of the IMF Managing Director. The ultimate objective of the IPSF is to scale up the mobilization of private capital towards environmentally sustainable investments.

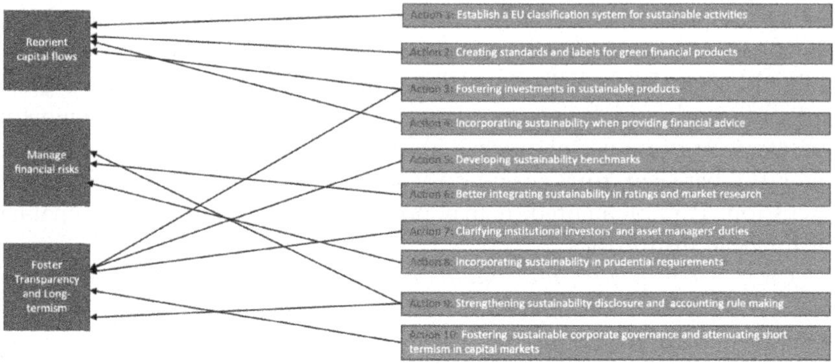

Fig. 3.4 Mapping the Action Plan on Financing Sustainable Growth against macro policy objectives (*Source* Author's elaboration)

This is a forum to strengthen international cooperation and coordination on approaches and initiatives for the capital markets (such as taxonomies, disclosures, standards and labels) that are fundamental for private investors to identify and seize environmentally sustainable investment opportunities. IPSS has also recognized the need to scale up sustainable finance beyond the European Union and harmonize and coordinate policies globally. Emissions in Europe are reducing both due to de-industrialization and to the positive impact of the implementation of measures intended to contain emissions. However, to have a positive effect global coordinated measured beyond Europe are essential.

In the communication to the European Parliament, the European Council, the Council, the European Economic and Social Committee and the Committee of the Regions of 11th December 2019 the European Commission outlined the terms of its commitment to tackling climate- and environmental-related challenges.

The document, commonly referred to as The Green Deal should be regarded as a further effort on the part of the Commission to fully implement the SDGs and to make Europe the first climate-neutral continent by 2050. The Green Deal significantly increases climate action and environmental policy ambition. It is not intended to cause a directional change in the European climate change policymaking. On the contrary its implementation is expected to advance existing policy objectives. The main objectives of the Green Deal and of its renewed sustainable finance strategy are twofold. First to design a set of deeply transformative policies. Old principles will continue to guide the actions of the Commission but will coexist with new concepts and new policy tools. The European Pillar of Social Rights will continue to occupy a central role in ensuring inclusion. For the first time, significant emphasis is placed on relatively new concepts such as energy poverty risk and the role of digital transformation as policy enabler. Energy poverty risk is described as the risk that households cannot afford key energy services or cannot afford the costs involved in increasing the energy efficiencies of their houses. Digitalization is regarded as a key tool for enabling climate change policies and monitoring their implementation and compliance. Digital solutions can help improve products data availability (origin, composition, history, etc.) and if combined with interoperable data and artificial intelligence solutions may facilitate evidence-based decisions and enhance policymakers ability to tackle climate change risk. Second, to pursue mainstream sustainability in all EU policies. As a result, we are expecting policymakers to support the creation of investment opportunities with a positive impact on sustainability for citizens, financial institutions and corporates. Furthermore, new regulations aimed at ensuring that climate, social and environmental risks are fully managed and integrated into financial institutions' risk management and the financial system as a whole are expected to come to light in the near future. In this respect, on 29th June 2020 The Climate Financial Risk Forum (CFRF) published a guide to help the financial industry address climate-related financial risks. CFRF was set up in March 2019 in the UK by the Financial Conduct Authority (FCA) to advance financial sector responses to the financial risks deriving from climate change.

The European Green Deal and Investment Plan and the Just Transition Mechanism are essential tools for the implementation of the Green Deal. The European Green Deal and Investment Plan is based on three pillars: (a) *financing*: it aims to mobilize public investments and help to unlock private funds through EU budget for an aggregate value of € 1 trillion before 2030; (b) *enabling*: this should include providing incentives to redirect public and private investments. The EU will provide additional tools for investors by putting sustainable finance at the heart of the financial system and will facilitate sustainable investment by public authorities by encouraging green budgeting and procurements and (c) *practical support*: The Commission will provide support to public authorities and project promoters in planning and designing and executing sustainable projects. The Just Transition Mechanism is a tool intended to ensure that the transition towards a climate-neutral economy happens in a fair way and it consists of three main sources of funding: (a) a Just Transition Fund: the fund is expected to be in a position to deploy up to € 7.5 billion in projects aimed at supporting the job market, investments in SMEs and start-ups incubators. In order to access the fund Member States will have to identify eligible territories through dedicated territorial just transition plans and match each euro made available by the fund with funds from the European Regional Development Fund and the European Social Fund Plus; (b) a dedicated just transition scheme under Invest EU which will be able to deploy up to € 45 billion. The scheme is intended to attract private investments in specific sectors including transport and sustainable energy; (c) a public sector loan facility with the European Investment Bank backed by the EU budget: the facility is expected to be able to commit up to € 30 billion in loans to the public sector in district heating networks and renovation of buildings.

On 4 March 2020 the Commission proposed a European Climate Law to turn the political commitment of climate neutrality by 2050 and more in general the goal set out in the Green Deal into a legally binding obligation. This act, which if implemented is expected to take the form of a regulation, will aim at achieving net zero emissions for EU countries as a whole mainly by cutting emissions, investing in green technologies and protecting the environment.

What Lays Ahead: Raising the Standards

The future regulatory landscape is expected to be characterized by more regulation aimed at defining and facilitating climate finance, more disclosure of climate-related risks and more stringent rules on the marketing of green investments to professional and retail investors. Prior to the 2020 pandemic, incentives for banks in the form of preferential prudential treatment intended to encourage lending to the green economy were also expected. As a matter of principle, in a risk-driven regulatory environment, the absence of supporting data should lead one to argue against such incentives. In practice, however, due to the current pandemic and general concerns on the asset quality of many large financial institutions in Europe and overseas discussing preferential prudential treatment appears a relatively futile exercise.

By increasing its climate policy ambitions, the Commission is triggering the need for more capital. The Commission estimates that reaching the current 2030 climate and energy targets alone would require investments of approximately € 260 billion by 2030. In addition, large segments of the working population will require upskilling and reskilling both of which are expected to require significant investments in education.

The Guidelines published by the European Central Bank setting out the supervisory expectations relating to risk management and disclosure represent a clear indicator of the changes that we may reasonably expect in the risk management and governance of banks.

The policy objectives, however, are extremely ambitious because of the relatively tight implementation timeframe and the costs involved, so there is a risk that they may not be fully implemented or that they will have to be adjusted based on the actual ability of the Climate Change Ecosystem to adapt.

Whether or not the full implementation of policy objectives is going to be delayed and provided that any delays remain within the boundaries of reasonability, like the Romantics, we should be comforted by the fact that the Climate Change Ecosystem seems to be embracing a sense of responsibility to all stakeholders and appears to be fully committed to inform and inspire all stakeholders to change and improve society (Forward, 2014) and, we should add, protect the planet.

References

Agreement Paris. (2015). *United Nations framework convention on climate change.*

Baker, M., Bergstresser, D., Serafeim, G., & Wurgler, J. (2018). *Financing the response to climate change: The pricing and ownership of US green bonds* (No. w25194). National Bureau of Economic Research.

Boehmer-Christiansen, S. (1996). Political pressure in the formation of scientific consensus. *Energy & Environment*, *7*(4), 365–375.

Climate Bonds Initiative. (2018a). *The climate bonds taxonomy. A guide to climate aligned assets & projects 2018.*

Climate Bonds Initiative. (2018b). *Green bonds policy: Highlights from 2017.*

Climate Bonds Initiative. (2018c). *Green bond highlights 2017.*

Climate Bonds Initiative. (2019a). *Green bond European investor survey.*

Climate Bonds Initiative. (2019b). *Green bonds market summary—Q3.*

Climate Bonds Initiative. (2019c). *Green bond pricing in the primary market: January-June 2019.*

Climate Bonds Initiative. (2019d). *Green bonds: The state of the market 2018.*
Climate Bonds Initiative. (2019e). *Green bond market highlights 2018.*
COM. (2016). *Communication from the Commission to the European Parliament, The Council, The European Economic and Social Committee and the Committee of the Regions: Next steps for a sustainable European future, European action for sustainability.*
COM. (2018). *Communication from the Commission to the European Parliament, the European Council, The Council, the European Central Bank, the European Economic and Social Committee and the Committee of the Regions; Action Plan: Financing Sustainable Growth.*
Ehlers, T., & Packer, F. (2017, September). Green bond finance and certification. *BIS Quarterly Review.*
European Commission. (2019). *Towards a sustainable Europe by 2030.*
Fatica, S., Panzica, R., & Rancan, M. (2019). *The pricing of green bonds: Are financial institutions special?* Available at SSRN. https://doi.org/10.2139/ssrn.3623146
Forward, S. (2014). *The romantics.* British Library.
Ghoul, S. E., Guedhami, O., Kwok, C. C. Y., & Mishra, D. R. (2011). Does corporate social responsibility affect the cost of capital? *Journal of Banking & Finance, 35*(9), 2388–2406.
Hachenberg, B., & Schiereck, D. (2018). Are green bonds priced differently from conventional bonds? *Journal of Asset Management, 19*(6), 371–383.
ICMA. (2018, June). *Green bond Principles voluntary process guidelines for issuing green bonds.*
Karpf, A., & Mandel, A. (2018). The changing value of the 'green' label on the US municipal bond market. *Nature Climate Change, 8*, 161–165.
Krechovská, M., & Procházková, P. T. (2014). Sustainability and its integration into corporate governance focusing in corporate performance management and reporting. *Procedia Engineering, 69*, 1144–1151.
Levillain, K., & Segrestin, B. (2019). From primacy to purpose commitment: How emerging profit with purpose corporations open new corporate governance avenues. *European Management Journal, 37*(5), 637–647.
Mathers, A., & Deonandan, R. (2018). Are the Sustainable Development Goals realistic and effective: A qualitative analysis of key informant opinions. *OIDA International Journal of Sustainable Development, 11*(03), 11–16.
Naciti, V. (2019). Corporate governance and board of directors: The effect of a board composition on firm sustainability performance. *Journal of Cleaner Production, 237*, 117727.
OECD. (2017). *Mobilising bond markets for a low-carbon transition.* Green Finance and Investment. OECD.

Wordsworth, W. (1835). *A guide through the district of lakes in the North of England: With a description of the scenery, &c., for the use of tourists and residents*. Hudson and Nicholson.

Zerbib, O. D. (2019). The effect of pro-environmental preferences on green bond prices: Evidence from green bond. *Journal of Banking & Finance, 98*, 39–60.

CHAPTER 4

SRI, ESG and Value of Sustainability

Laura Pellegrini

Introduction

This chapter approaches some of the most relevant "new frontiers issues" of corporate governance and finance issues: CSR, ESG indexes and the value of sustainability.[1]

Although corporate finance has historically researched about the determinants of stock and bonds returns and modelling future yields, recently the corporate governance has focussed its attention on measuring the impact of non-financial information on listed companies' corporate financial performance. According to the efficient market theory, all new information has the potential to impact the market value of shares. Therefore, it could be stated that the more complete and more reliable the information available, the more accurate is the valuation of the

[1] I would like to thank Niketa Mehmeti and Giulio Squarta for some interesting insights.

L. Pellegrini (✉)
Department of Economics and Finance, University of Bari, "Aldo Moro", Bari, Italy
e-mail: laura.pellegrini@uniba.it

© The Author(s), under exclusive license to Springer Nature Switzerland AG 2022

C. B. Pellegrini et al. (eds.), *Climate Change Adaptation, Governance and New Issues of Value*, Palgrave Studies in Impact Finance,
https://doi.org/10.1007/978-3-030-90115-8_4

future performance of equity. Even if extra-financial information may not necessarily affect the price of a company's share during normal operations, in cases where reputational or monetarily quantifiable litigation risk exists, investment professionals pay strong attention to the respective information.[2]

This field of study has become more relevant over time due to the increasing attention of investors for these issues. Needless to say, the challenge is to verify whether considering sustainability, environmental and social issues also payoffs in terms of performance and added value to the firm.

It is however evident that evaluation of ESG matters enables a thorough understanding of the risks and opportunities a company faces, allowing enhanced security selection and risk management. Additionally, ESG analysis leads to an improved understanding of how future trends could affect a certain industry or the whole economic landscape for that matter.[3] In particular, the ESG scores combine different elements,[4] among which climate change—as one of the most prominent and challenging environmental issues facing companies—has a particular relevance for financial markets. It is foreseeable that companies will need to consider various features and conditions in the near future and require to operate with a very-close boundaries in terms of sustainability to survive at first and expand later.

What We Know About Corporate Social Responsibility

Through the years, corporations have achieved a dominant position within society, providing employment, generating wealth and of course consuming natural resources. Moreover, companies play key roles in public policy especially in countries where the standard of governance

[2] That is why companies make an increasing effort to provide investors with disclosures on extra-financial aspects which capture additional dimensions of corporate performance that are not accounted for within financial data.

[3] Financial professionals, for instance, anticipate that ESG issues and climate change in particular will "gradually but powerfully change the economic landscape (...) cause periodic sharp movements in asset prices" (Llewellyn, 2007).

[4] Environmental, social and governance score.

is poor or in cases where internal governance mechanisms are inadequate. Generally, responsibilities rise from certain rights granted. Given the increased power, pervasive presence and large resources over which they have control, companies find themselves coming under pressure to act in a socially responsible way.

In defiance of past and current problems, capital markets are demanding greater transparency and greater attention on corporate governance issues. Climate change and financial crisis are two strong examples of the disastrous impact that negligence of corporate responsibility issues did cause. From this point of view corporate influence is so strong and pervasive. Hence, the public, who is the ultimate stakeholder, is demanding greater diligence for a wide range of issues including environmental pollution, consumer rights, employee relations, human rights and corruption.

Over recent decades, the practise of Corporate Social Responsibility has become part of mainstream corporate activity with virtually more than 85% of S&P500 companies including it on their agenda in some form, thus demonstrating a growing recognition of the great interest expressed by investors. Often confused with other notions like sustainability, business ethics or company value, the term "CSR" has lacked an ultimate definition and not rarely has been redefined by its practitioners, most notably companies, to suit their own viewpoints and purposes. The main reason is to be found in the core commitment for a company. Being socially responsible means not only fulfilling legal obligations and creating value but it goes further, considering compliance and investing in human capital, environment issues and relations with stakeholders.

As the world has acknowledged enormous changes in the technology, it has also witnessed a growing crisis regarding climate change, resource reduction and discrepancy in wealth distribution, corruption and violation of human rights. There is a game-changing campaign focussed on engaging industry, policymakers and academics in the debate concerning the role of corporations in society.

Some of the well-known neoclassical economics and several management theories based on the idea that the corporation's objective is profit maximization subject to capacity constraints and the central role of the ultimate shareholder as residual claimant (Jensen & Meckling, 1976; Zingales, 2000) are changing. The path that is taking shape looks very different today.

As it usually happens when some recent innovation begins its process of diffusion and institutionalization, univocally defining the terms and concepts surrounding sustainable investing is, for the time being, particularly complex. This is because, while a lot of effort has been recently put in by academics, practitioners and regulators to clarify the taxonomy around the topic, the task is far from being completed. In order to better understand the investment-related side of the subject, it is more useful to firstly approach the issue from a business perspective, given that its process of diffusion initiated much earlier and the terms involved are generally more widely accepted.

The term Corporate Social Responsibility (CSR) was firstly holistically defined by Carroll (1999) in terms of four responsibilities that a company has: an *economic responsibility* to produce goods/services and to be profitable, a *legal responsibility* to follow and abide the law, an *ethical responsibility* to follow "additional behaviours (...) that are not necessarily codified into the law but nevertheless are expected of business by society's members" (ibid.) and a *discretionary responsibility* to contribute towards other social purposes (e.g. philanthropy).

Through time, other notions refined and expanded the theme of CSR. Among those, some of the most relevant are the concept of stakeholders and stakeholder management (Freeman, 1984), according to which a business should consider a broader set of agents other than solely shareholders, and the idea of a *triple bottom line* (Elkington, 1997) where the performance of a company should be assessed on profit, people and the environment, so as to consider the full cost of doing business.

More recently, a new broadly accepted definition was provided by Sheehy (2015), according to which CSR is "a *type of international private law and can be defined as a socio-political movement which generates private self-regulatory initiatives, incorporating public and private international law norms seeking to ameliorate and mitigate the social harms of and to promote public good by industrial organisations*". As it is clear from both definitions, the themes surrounding CSR encompass many different academic disciplines, which may lead to slightly different frameworks being applied. In that sense, according to Sheehy (2015), while the economic discipline considers CSR through the lens of the theory of the firm, agency theory and market failure, the business scholarship focusses on moral obligation, sustainability, licence to operate and reputation.

Given that the disclosure of CSR-related information is still largely discretional, while there is some evidence for specific sectors and regions, it is not quite possible to outline various relationships between disclosure and a firm's characteristics. That said, generally speaking, while larger corporations (by number of employees) seem to be more likely to communicate their CSR practises but fail to implement it (i.e. implementation gap), smaller firms better integrate CSR decisions but face problems disclosing their actions (i.e. communication gap) (Baumann-Pauly et al., 2013; Wickert et al., 2016). This is because, according to Fassin (2008), large firms face high costs while implementing CSR practises across their operations but have comparatively low expenses in preparing CSR reports; conversely, smaller companies face smaller implementation issues but higher reporting costs.

The concept of social responsibility is not so easy to define. Evidences of concerns regarding the impact of industrialization on society are traceable since the Victorian era but the formal writing on social responsibility is largely a product of the twentieth century. Both the positive and negative impacts that business had on society generated a wide public, political and academic debate. The 1950s marked the advent of contemporary corporate responsibility, distinguishing it from corporate philanthropy, as corporations today are engaging in programmes which, rather than just giving back to the community, affect core management practises.

In the early phase of the CSR debate, it was more common to refer as social responsibility rather than corporate social responsibility. Initially it began with a focus on the role of business leaders, particularly on how they managed their companies with a view to society and how they gave back to their local communities. Starting from the mid of twentieth century, the focus shifted towards the behaviour of companies rather that of individuals (Blowfield & Murray, 2014). The pioneering work by Bowen (1953) gives an initial definition of the SR of businessmen as: "*It refers to the obligation of businessmen to pursue those policies, to make those decisions, or to follow those lines of actions which are desirable in terms of the objectives and values of our society*".[5]

A further attempt to formalize more accurately what CSR means is proposed by Davis (1973) arguing that SR should be seen in a managerial context, and remarking that the Socially Responsible business decisions could be justified by a process of reasoning as having a good chance of

[5] Bowen's book and definition represented the most notable literature from the 1950s, and it is easy to see how it marks the modern, serious discussion of the topic. (about the evolution of CSR see also Carroll (1999).

bringing long-run economic gain to the firm, thus paying it back for its socially responsible outlook.[6]

But, there were also contrary opinions about the issue. Milton Friedman (1970) from this point of view stated that only people have responsibilities. A corporation is an artificial person and its business as a whole cannot claim to have responsibilities, even in the vague sense.[7] What happened? Is something in the rules of the game changed?

Why Should We Care About Social Responsibility Issues?

We are living in a world where various resources have started to be in short supply due to not only the limited availability of natural resources but also human-caused crises accompanied by misallocation of resources are causing significant unsustainable conditions on the planet.

The 2008 global financial crisis has caused serious damage worldwide by emphasizing how the financial system has been working in a way that is unsustainable. Furthermore, the financial crisis pointed out that the conventional paradigm in finance developed since the 1950s is unable to manage with the problems of financial systems, financial markets and behaviour of financial institutions, and failed to understand the suitable role of finance in society and the economic system as a whole. Putting aside what triggered the recent crisis, an important question would be regarding the sustainability of the world's systems and economies going forward and what would be the way out. The financial crisis has led many *institutions* and research groups around the world to reflect and engage in debates regarding the key issues of modern finance, the impact of finance on society and the sustainability of the financial system.

[6] "It is the firm's obligation to evaluate in its decision-making process the effects of its decisions on the external social system in a manner that will accomplish social benefits along with the traditional economic gains which the firm seeks. It means that social responsibility begins where the law ends. A firm is not being socially responsible if it merely complies with the minimum requirements of the law, because this is what any good citizen would do. Social responsibility goes one step further. It is a firm's acceptance of a social obligation beyond the requirements of the law".

[7] "There is one and only one social responsibility of business—to use its resources and engage in activities designed to increase its profits so long as it stays within the rules of the game, which is to say, engages in open and free competition without deception or fraud."

Something of our well-known conventional paradigm is changing. Moving from Capital Asset Pricing Model (Lintner, 1965; Mossin, 1966; Sharpe, 1964) and the trade-off between risk and return, together with the Markowitz's portfolio theory (Markowitz, 1952), Modigliani and Miller's arbitrage principles (Modigliani & Miller, 1958, 1963) and Black-Scholes-Merton pricing (Black & Scholes, 1973; Merton, 1973) the new modern financial paradigm focusses on assumptions such as market efficiency and rational behaviour. Modern scientific finance, based on neoclassical economies, constructs an ideal, perfect and abstract reality of capital markets and market participants, which notably diverges from the real world where the norm is that markets are not perfect and where economic actors behave in an irrational way. For almost a century, however, CSR and conducting business by respecting ethical principles have become a public concern. In the financial world, Responsible Investing represents one of the first movements attempting to question whether financial markets are appropriately serving the society within which they operate.

Various visions about CSR have emerged during the years. Among others, the interpretation proposed by Davis (1973) deserves a discussion due to its relevance and foresight over the years. The need to consider socially responsible behaviour into the operational framework of corporations, arise with the evolution of the CSR concept along time.

At first, we need to pay attention to SR behaviour because (1) *it is in the long-run self-interest*. The rationale of this argument has its roots in the assumption that society expects business to accomplish social good, and corporations must fulfil these expectations in order to generate a profit in the long run. Businesses that are sensitive to community's needs have greater opportunities to conduct their activities in better social environment. The responsiveness towards community will result in the reduction of turnover and absentee-ism. Recruiting and quality of labour will ameliorate. Improvements in the welfare will lead versus the decrease of criminality and as a consequence, less money will be spent for protection and fewer taxes will be paid to support the defence forces. Although it is as much a sophisticated concept as contradictory, spending money for social programmes will actually lead to higher profits for the business. The second point concerns the idea of a good *Reputation (2)*. Each firm, during its life, try to maintain a well-established reputation. By doing good and giving back to the community, the firm enhances its public image aiming to gain more customers, better employees and greater exposure to investors. This aim needs to be considered as an "opportunity value". Closely related to the idea of reputation is the aspiration to carry on the *viability of the business* (3) in the long run. Society represents the demand side and it essentially provides business with its charter to exit. In order to keep their power and social role, firms should comply and respond to the needs of the society. Contrary to the public image that serves the stand-alone firm, the concept of viabilities benefits the whole business system.

Again, one of the reasons why a corporation embraces a socially responsible behaviour is driven by its attempts to *avoid regulation by the government (4)*. Regulation implies compliance costs for a business and in the same time it limits its flexibility regarding decision-making. Consistent with political philosophy to maintain the power decentralized for democracy's sake, businesses are motivated to hold on to their freedom in decision-making as it permits them to choose the degree of responsiveness to market and social factors. By voluntarily engaging in socially responsible behaviour and doing contemporarily a public and a private good, an economic entity can prevent government intervention.

Firms pay so much attention to *cultural norms (5)*, as a strong determinant of behaviour. As any other individual, a business leader's behaviour is constrained by a set of cultural patterns. Sociocultural norms in a business play the same decisive factor as legal, technical and market restrictions do. Though its decisions are not completely affected by these social constraints, it's in the business's interest to be sensitive to the changes that are happening in the social sphere and make sure to be reactive to the new demand side.

It has been shown that it is in the best *interest of the stockholder (6)* to get involved in certain kind of responsible behaviour. In the case of a single firm investor, its profits will largely depend on the degree of openness such firm has on the responsibility behaviour. Meanwhile, stockholders of a diversified portfolio have more chances to see their account grow.

As we already know, social responsibility occupies a relevant place in the agenda of firms and is gaining momentum, given its role as a primary driver for strategic product and business ***model innovation (7)*** (Davis, 1973[8]).

We have so many evidences of direct monetary cost related to scandals, crisis and so on.

Let us think about Madoff scandal, the $13 billion fine paid by JP Morgan due to its role in the sub-prime lending crisis or the British Petroleum's oil spill in the Gulf of Mexico marking a cost of around $42 billion only for clean-up and compensations.[9] We know that the cost in dollar terms is easily quantifiable. But, how can we estimate the social impact of these events over everyday life? Delaying the process of pursuing social responsibility strategies, businesses may find themselves constantly occupied with putting out social fires and as a consequence, slowing down their primary goal achievements such as producing goods and services. Since these problems must be dealt with at some time, it is far more convenient for the business to take all the necessary precautionary measures to limit such impact and the spill over effects deriving from it. ***Because prevention is better than curing (8).***

Emerging Approaches on ESG Issues and Sustainability Report: The State of the Art

As the interest in ESG data has steadily increased in the recent past, this growing demand has not been met by a satisfactory common standard in disclosing and reporting non-financial information. While many possible reasons behind the phenomenon exist, these can be summarized in two broad issues. The first one concerns the novelty and complexity of the elements surrounding CSR that made it difficult for regulators around the globe to timely and uniformly provide legal frameworks, either mandatory or recommendation-based, through which corporations could receive clear standards on the matter. Moreover, companies were faced with the similar concern of being unable to clearly understand which information to disclose, what was to be considered materially relevant for investors and the general public, which reporting standards to follow if

[8] He argued that institutions that work in the social areas seem to be particularly lacking management talent, and business is known worldwide for its investment in this field and as consequence, its innovative ability. From this point of view the collaboration between these two types of institutions is beneficial for both parties.

[9] Estimation made by *The Economist*.

deciding to disclose and if so, under which tool (i.e. separate CSR report, include ESG data in financial statements or integrated reporting). The aim of this section is to present a brief overview of the issues related to sustainability reporting, the voluntary reporting frameworks currently available, the growing role of mandatory disclosure and some recent developments in the regulatory landscape.

The first examples of voluntary sustainability reporting can be traced back to the early 1960s in both the USA and Europe, mostly driven by the social pressure from NGOs and a growing awareness of society towards ethical themes. This trend continued in the following decades, expanding in both relevance and diffusion but without any type of widespread convergence towards shared best practises on how or what to disclose.

The concept of acting in a responsible way in finance has taken different patterns through the years. From the beginning of the new millennia, greater attention to the ESG issues and sustainability has been paid by mainstream investment institutions.

For a long time, the Environmental, Social and Governance (ESG) considerations have been associated with "Responsible Investors", a marginal and morally conscious group of investors that in addition to the traditional financial criteria, used information related to environmental, social and governance performance to screen and choose their investments.

It is important to emphasize the fact that the integration of ESG information in the investment decision-making process of companies follows a rudimentary different approach compared to Responsible Investing (RI). While RI is the outcome of moral and ethical concerns, the new trend on ESG matters and integration of ESG information into traditional investing activities is mainly business driven.[10] Responsible Investing gained relevance into the financial world during the 1990s, when it became a new trend in investing and many financial institutions started to develop RI funds to respond rapidly to the increasing market demand for RI. Even though the implementation of ESG considerations started to become relevant, ESG continued to remain disconnected from the traditional investment activities. Up to the last decades. This will be a *turning point*

[10] The genesis of RI is closely related to faith-based organizations. Just think that the first RI initiatives were undertaken by religious organizations that didn't want to finance and be associated with certain activities or companies (the so-called "sin stock" usually referring to companies that supplied for instance alcohol, tobacco, gambling and weapons).

since an increasing number of initiatives aiming to integrate ESG issues into mainstream investment and into valuation process came to life.

In response to the growing demand for disclosure, the European Union made a step forward with the adoption of the Non-Financial Reporting Directive in 2014, emphasizing the increasing importance of non-financial information, therefore deciding to legislate what is expected in terms of transparency. The new directive's scope is to help investors, consumers, policymakers and stakeholders to assess the non-financial performance of large companies and at the same time to encourage these companies to adopt a more sustainable and responsible approach in conducting business.[11] More recently, a continued interest in ESG analysis from both investors and stakeholders, pushed legislators in a growing number of jurisdictions to begin mandating some form of disclosure, even though with varying degrees of stringency.

As discussed in the previous chapters, sustainability disclosure requirements apply to a relatively limited number of market participants. While this may not be a problem per se—it could facilitate companies in the decision of providing some information rather than none—voluntary reporting comes with a number of issues. This is because, while investors require information that is timely, reliable, consistent and comparable across companies and time (Ho, 2017), voluntary disclosure, especially when there is a lack of a market-wide reporting framework, could fail in meeting these criteria, resulting in a confusing scenario for both corporations and investors. Additionally, this may lead to CSR reports that are costly and cumbersome to produce for companies but lack the financially material information that investors actually care about. In that sense, voluntary reports may have limited use for investment analysis purposes: reports have historically focussed on qualitative data, difficult to analyse when compared to quantitative metrics; the reliability and credibility of financial reports are also based on the revisory role of auditors, but only a small fraction of CSR reports undergoes an auditing process; the timing of voluntary reports is often on a different time frame with respect to traditional financial statements, making it complex for analysts to integrate the two (CFA Institute, 2015).

[11] Non-financial reporting supports business in various ways. Some of these benefits could be classified as (i) Reassurance of stakeholders' trust, to transform risks into opportunities: (ii) Learning by doing, creating value; (iii) Continuous improvement, bring more positive impact; (iv) Enhancing integrity, increasing transparency.

Several international standards and initiatives are suggested and have been used in the context of reporting:

1. Global Reporting Initiative (GRI) The most widely adopted framework of sustainability reporting has been the GRI Sustainability Reporting Framework, with the aim to standardize CSR reporting so as to achieve a level of credibility and comparability similar to that of the more common financial statements. It consists of a collection of reporting guidance documents designed to assist companies in preparing ESG disclosure and sustainability reports. Launched in 1997 by CERES[12] and UNEP[13] with the scope of establishing and further developing a guide in how to report for the "Triple Bottom Line" which accounts for economic, environmental and social performance by corporations, has the final goal to gain the gradual convergence of sustainability reporting with financial reporting in terms of credibility, rigour and comparability.
2. Integrated Reporting Framework (IIRC) aims to provide organizations with a framework for integrated reporting on strategy, financial position and performance. The Integrated Reporting Framework considers both financial and non-financial aspects by assessing six forms of capital.[14] The principles of integrated reporting provide a framework that better notifies investors and stakeholders with information on how an organization can create value concentrating on social, environmental and economic domains. The topics discussed in the new EU Directive are linked to the various forms of capital included in the framework making the management report meaningful and at the same time, providing insights on how the *three bottom-line* themes are part of the value creation process.
3. OECD guidelines, which provide advices on how companies deal with matters such as supply chain responsibility, environment, human rights, corruption and employment and labour relations.
4. Global Compact. To this point, the United Nations seeks to bring together several parties like companies, UN organizations, trade unions and civil society organizations in order to further contribute

[12] Coalition for Environmentally Responsible Economies.

[13] United Nations Environment Program.

[14] Financial, manufactured, human, intellectual, social and relationship and natural.

to international corporate social responsibility. It has the intention to achieve this goal through the Global Compact initiative, endorsed by more than 9,500 companies in over 160 countries, which addresses the same topics as the Non-Financial Directive, including different issues concerning human rights, working conditions, environmental protection and the fight against corruption and bribery.

5. Similarly, to the GRI, the Sustainability Accounting Standards Board (SASB), founded in 2011, is an organization aimed at developing standards for the disclosure of relevant sustainability information. Differently from the GRI Standards, SASB Standards are intended to specifically highlight sustainability matters that are financially material for investment decisions. In that sense, SASB provides standards that are specific for 11 sectors and 77 industries, allowing companies to focus on those issues that are most likely material for their investors. As a consequence, these standards tend to analyse topics much more granularly and in a quantitatively measurable way.[15]

The deadline for the EU Member States to adapt into their national legislation the Directive was December 2016 and starting from 2017 large companies, both in terms of size and socio-economic impact, were expected to disclose the mandatory reports on non-financial information. As the EU Member States complete the national transpositions of the Directive, Europe is setting a clear path towards greater business transparency and accountability on social and environmental issues. The new laws are expected to play a crucial role in incentivizing and concretizing private sector's action and commitment towards a more sustainable agenda. Moreover, the practise of reporting can help enhancing the responsible business conduct of the companies and make a step forward to a more sustainable future rather than merely complying with the legal requirements. Not to be neglected is the fact that there are some differences in reporting landscapes across States. The EU leaves some discretion to the Member States, albeit complying with certain parameters

[15] SASB has identified a total of 26 sustainability related issues which declined for each industry as both disclosure topics and accounting metrics.

concerning the implementation of the new legislation into their national law and according to the different country's framework.[16]

Most of the initiatives that provide frameworks on sustainability disclosure are predominantly NGOs driven, in some way as a direct consequence of how the sustainable finance movement initially started. As an exception to this trend, the Task Force on Climate-related Financial Disclosures (TCFD), established in 2015 by the Financial Stability Board, is one of the few industry-led initiatives. The TCFD was initially established with the intent of providing practical recommendations on climate-related financial risk disclosures to both companies, helping them understand what financial markets want from voluntary reporting, as well as investors, lenders and insurers on how to better assess, price and manage their short, medium- and long-term exposures to climate risks. In that sense, the role of the TCFD has been to bring climate-related financial reporting to a mainstream audience (TCFD, 2020).

Although the implementation of the new legislation is at its early phase, the majority of Member States had already experienced dealing with ESG themes. Such fact holds as an assurance for future development in the integration process and sophistication in the cautionary measures undertaken towards a more sustainable future.

What We Mean for Sustainable Investing and How to Promote It

Having introduced what CSR is and how it is gauged via ESG data, it is now possible to focus on the various investment approaches related to the theme of sustainability. There is currently a huge number of definitions and investment styles that consider ESG issues. In most cases, these elements lack a formal definition and are often used interchangeably (PRI, 2020). There are a lot of different definitions and approaches of sustainable investment. Among the others, some of the most important can be summarized as follows.

At the most inclusive level, ***sustainable investing*** is defined by the Global Sustainable Investment Alliance (GSIA) as "an investment

[16] See, for example, Denmark and Sweden, or again France and Italy evidences (Jeffwitz and Gregor, 2017), Comparing the implementation of the EU Non-Financial Reporting directive in the UK, Germany, France and Italy., Frank Bold Organization EU Directive disclosure of non-financial information and diversity information—EY (2017).

approach that considers environmental, social and governance (ESG) factors in portfolio selection and management" (GSIA, 2018). This definition is purposely broad so as to encompass the widest range of investment strategies.

Moving to a more detailed definition, the Principles for Responsible Investment (PRI) define *responsible investment "as a strategy and practice to incorporate environmental, social and governance (ESG) factors in investment decisions and active ownership" (PRI 2020a). Differently from other approaches that directly target specific ethical or moral objectives, responsible investment "can and should also be pursued by the investor whose sole focus is financial performance"* (ibid.).

Finally, socially responsible investment or, more commonly, ***sustainable and responsible investment*** (SRI) is defined by the European Sustainable Investment Forum as*: A long-term oriented investment approach which integrates ESG factors in the research, analysis and selection process of securities within an investment portfolio. It combines fundamental analysis and engagement with an evaluation of ESG factors in order to better capture long term returns for investors, and to benefit society by influencing the behaviour of companies* (Eurosif, 2018).

To a careful reader, however, some peculiarities cannot be avoided. While these three definitions share some common elements, there are some slight differences. For example, an investor having the sole objective of maximizing risk-adjusted returns could employ responsible investing by implementing ESG data to enhance risk management and compliance processes; however, that same investor may not be interested in SRI since it would require a direct intention to generate positive externalities.

It is important to note that, for the time being, such classifications do not constitute any legal or otherwise strict taxonomy requirement, but rather are an attempt from organizations and academics to guide the investor community through the various options available.

The recent surge in interest towards sustainable investing can be traced back, making some simplifications, to mainly two factors. On the one hand, society's generalized increase in awareness to global challenges (e.g. climate change, inequality, etc.) and on the other due to the continuous efforts made by national and international organizations promoting sustainable investing and providing the tools necessary to correctly understand the theme and guiding its implementation. The role of such organizations is particularly relevant in this fairly novel area of finance

Table 4.1 Principles for Responsible Investment

Principle 1	Principle 2	Principle 3	Principle 4	Principle 5	Principle 6
Incorporate ESG issues into investment analysis and decision-making processes	Be an active owner and incorporate ESG issues into ownership policies and practices	Seek appropriate disclosure on ESG issues by the entities in which invest	Promote acceptance and implementation of th Principles within the investment industry.	Work together to enhance the effectiveness in implementing the Principles	Report on activities and progress towards implementing the Principles

Source Principles for Responsible Investment (2020) "PRI Brochure"

since national and international legislators have generally shown limited effort in creating legal frameworks for the industry.

Among the many examples of such institutions,[17] I would like to focus on the most relevant associations based on size and importance. Additionally, a focus is made on the leading institutions for the Italian and UK market.

At first, the Principles for Responsible Investment (PRI) is the largest promoter of responsible investing, and it's actively involved in supporting its signatories in understanding the implications of ESG issues, facilitating their implementation throughout the investment process and across asset classes, providing research and professional training, supporting practitioners and regulators in overcoming obstacles for a broader diffusion in the industry (Table 4.1).

The PRI was, then, officially launched in 2006 with 64 founding signatories and it currently accounts for more than 3,000 signatories—both asset owners and asset managers—overseeing in excess of $100 trillion of AUM (ibid.), as highlighted in Fig. 4.1.

In a Worldwide framework, it is important to mention the Global Sustainable Investment Alliance, the largest network of national and

[17] One of the most influential early collaborative projects developed, aiming to promote the integration of ESG information into mainstream research, stands the Enhanced Analytics Initiative (EAI). The former is an international collaboration between asset owners and asset managers launched at the end of 2003 to incentive researchers to take account of the impact of non-financial issues on long-term investment.

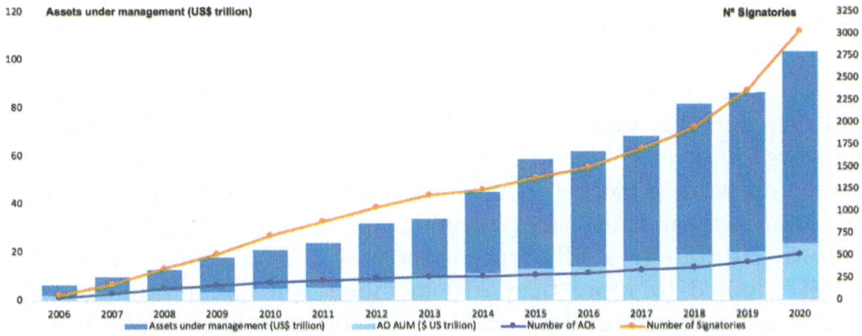

Fig. 4.1 PRI signatory and AUM growth (*Source* Principles for Responsible Investment [2020] "What is Responsible Investment")

international membership-based sustainable investment organizations.[18] Formally active from 2012, the role of GSIA is mainly addressed to coordinate the actions of its members, increasing the global visibility of sustainable investing and producing periodic surveys and reports. Specifically, the organization produces a biannual global report particularly useful for tracking market trends and comparing the specific features between regional markets.

At the European level, as previously introduced, the Eurosif is the leading regional association promoting the growth of SRI. Constituted by the national sustainable investment forums (SIFs) of Austria-Germany-Switzerland (FNG), Italy (FFS), France (FIR), Spain (Spainsif), Netherlands (VBDO) and UK (UKSIF), its role is to promote the advancement of SRI in the European marketplace. In that sense, Eurosif shares and disseminates best practises, promotes the collaborations between national forums, acts as a lobbying agent at the supervisory and legislative level, produces and disseminates research on SRI—including market trend reports.

[18] Among the members we found: Eurosif, the United States Sustainable Investment Forum (USSIF), the Responsible Investment Association Australasia (RIAA), the Responsible Investment Association Canada (RIA Canada), the UK Sustainable Investment and Finance Association (UKSIF), the Dutch Association of Investors for Sustainable Development (VBDO) and the Japan Sustainable Investment Forum (JSIF).

If you consider our Italian domestic context, the "Forum per la Finanza Sostenibile" (FFS or ItaSIF), founded in 2001, is the association promoting SRI for the Italian marketplace. Currently accounting for 111 members (i.e. asset managers, banks, pension funds, NGOs, etc.), its activities are divided into research (analysing best practises and tracking market growth), projects (organizing campaigns and conferences to raise SRI awareness) and advocacy (engaging with national and European institutions to promote a SRI legal framework). Of a similar nature, the UKSIF, joined by more than 240 members, aims at informing, influencing and connecting all the relevant stakeholders (financial institutions, policymakers and the public) in the process of fostering the growth of the UK SRI market. As with other SIFs, the organization fosters engagement, facilitates discussion and produces research and reports.

ESG Scores and Valuation Metrics

When potential investors are about to evaluate a company's performance, financial information is always their first consideration. But this is a historical idea of source value. Recently, environmental, social and governance efforts are also increasingly being considered. ESG factors cover a wide spectrum of issues that traditionally were not part of financial analysis, yet now have gained financial relevance. This might include how corporations respond to climate change, how good they are with water management, how effective their health and safety policies are in the protection against accidents, how they manage their supply chains, how they treat their workers and whether they have a corporate culture that builds trust and fosters innovation.[19]

In this sense it should be clear that, differently from other traditional financial metrics, measuring a firm's performance and stance towards these themes concerning CSR and SRI may be complex. This is partly due to the inherent nature of the thematic but also to the lack of a unique measurement approach.

Generally speaking, the non-financial information covering the CSR aspects of a company's performance is considered as Environmental Social and Governance (ESG) data. In that sense, ESG information can be

[19] The remarkable rise of ESG Investing—Georg Kell, *Forbes*, 11 July 2018.

used to track, compare and analyse how different companies behave with respect to ethical and sustainability-related issues.

As with CSR, it is not possible to precisely define all the different aspects covered in each category. An overview of the most common issues related to ESG could be summarize at first in Fig. 4.2.

Differently from standard financial metrics which are fairly straightforward to measure and internationally defined by accounting standards, identifying and measuring ESG issues is more challenging. Nevertheless, the ever-growing demand for reliable ESG data has pushed a large number of financial data providers to broaden and update their services. Unfortunately, the ESG metrics used differ greatly both in terms of application—e.g. sustainability indices, portfolio ratings, company reports, etc. (Eccles & Stroehle, 2018)—and in terms of the actual indicators being analysed and the methodology used (Chatterji et al., 2016). This confusion may also lead to different data providers assessing the same issue for the same company in different ways (Eccles & Stroehle, 2018).

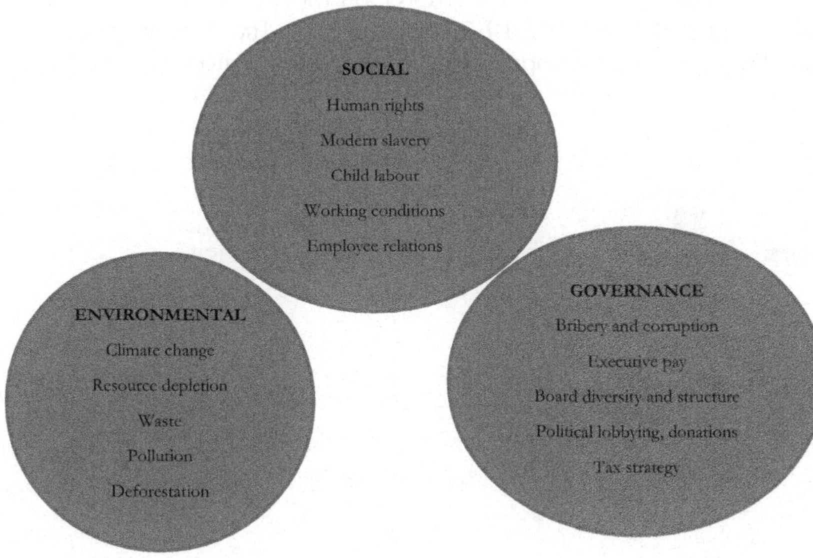

Fig. 4.2 ESG issues, categories idea (*Source* Principles for Responsible Investment [2020], "What is Responsible Investment")

While the first attempts to provide ESG data can be traced back to the 1970s (Friede et al., 2015), as of recent, the Global Initiative for Sustainability Ratings (GISR) estimates that more than 100 providers are collecting, analysing and distributing ESG data. Such marketplace is quite heterogeneous with both for profit and not-profit organizations, some focussed on specific issues (e.g. climate change) while others cover broader topics and with some companies which solely provide data and research whereas others also offer consulting and portfolio technology solutions (Eccles & Stroehle, 2018). As it can be seen from Fig. 4.3, the sector has been recently faced with vast consolidation. In most cases, large traditional data providers (e.g. Morningstar, Thomson Reuters, Bloomberg, MSCI, etc.) decided to acquire or invest heavily in smaller and boutique ESG-focussed providers, rather than forming such solutions internally.

Beside the growth in data providers and their related offerings, a large amount of sustainability indices for a number of regions and asset classes has been launched. Among those, some of the most relevant are the Dow Jones Sustainability Index, the FTSE4Good Index, the MSCI World ESG Index and the Barclays MSCI ESG Fixed Income Index. As it can be seen from Fig. 4.4, the development of sustainability indices has been fairly recent, and the market is expected to continue expanding.

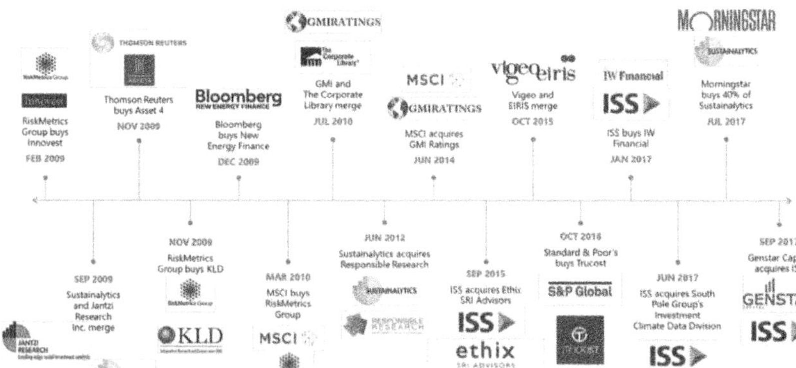

Fig. 4.3 Recent ESG data providers' consolidation (*Source* BrownFlynn [2018] "The ESG Ecosystem: Understanding the Dynamics of the Sustainability Ratings & Rankings Landscape February")

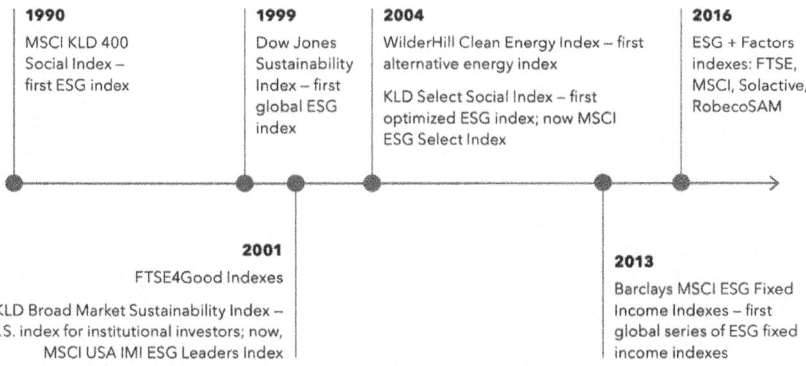

Fig. 4.4 Origin of ESG indices (*Source* iShares [2019] "An Evolution in ESG Indexing")

Nowadays, ESG investing is estimated to be around a quarter of all professionally managed assets around the world, and its rapid growth builds on the Socially Responsible Investment (SRI) movement that has been around much longer. But unlike SRI, which is based on ethical and moral criteria and uses mostly negative screens, such as not investing in alcohol, tobacco or firearms, ESG investing assumes that ESG factors have financial relevance.[20]

Across all industries, many investors try to understand companies' relative performance in the relevant ESG areas to understand how both, opportunity and risk are managed, and to determine how ESG components help companies drive success factors like growth, innovation and talent retention. ESG ratings provided by specialized rating institutions play an important role in the decision-making process of managers and investors who care about social responsibility.

The ESG issues are gaining momentum and greater attention each day. Investors are including sustainability considerations into their decisions; rating agencies are implementing ESG performance indicators in addition to financial indicators when revising the creditworthiness of issuers, and banks are lending at advantageous rates to companies that are largely involved in pro-sustainability matters. In other words, the route towards the future of financial system comprehends the incorporation of

[20] The remarkable rise of ESG—Georg Kell, *Forbes*, 11 July 2018.

sustainability in the mainstream behaviour of economic actors in every stand.

As we already said before, there are several providers of ESG ratings. Typically, an ESG score or rating is arrived at by measuring a company's performance against pre-determined metrics that may differ from a provider to another and are an integral part of the methodology used by each provider.[21]

Focus on Thomson Reuters buys Asset4 data providers for ESG rating we can approach to the ESG Combined Scores. This is a very commonly used rating for empirical approaches.

According to Thomson Reuters buys Asset4 definition, this score "*provide a rounded and comprehensive scoring of a company's ESG performance based on the reported information in the ESG pillars with ESG controversies overlay captured from global media sources. The main objective of this score is to discount the ESG performance score based on negative media stories*". Furthermore, "*Thomson Reuters captures and calculates over 400 company- level ESG measures, of which we have carefully selected a subset of 178 most comparable and relevant fields to power the overall company assessment and scoring process. The underlying measures are based on considerations around comparability, data availability, and industry relevance. They are grouped into 10 categories. A combination of the 10 categories, weighted proportionately to the count of measures within each category formulates the final ESG Score, which is a reflection of the company's ESG performance, commitment and effectiveness based on publicly reported information*".

Thomson Reuters provides a rounded and comprehensive valuation of the ESG performance of companies by analysing over 400 firm-level ESG measures, of which it selects a subset of 178 most comparable and relevant fields to power the overall company assessment and scoring process. The underlying measures are based on considerations around comparability, data availability and industry relevance (Fig. 4.5).

[21] However, in most cases each rating metric is a sub-category of one of three macro categories: environmental, social and governance. For completeness, it should be noted that some European practitioners are using the term ESG rating when referring to the assessment process which has to be conducted in order to establish whether or not a particular bond complies with the Taxonomy or the Green Bond Standards. In this respect, due to the nature of the process it may appear more appropriate and coherent with the expected forthcoming regulation and the proposal made by the TEG to classify the process as a verification rather than a fully fledged rating process.

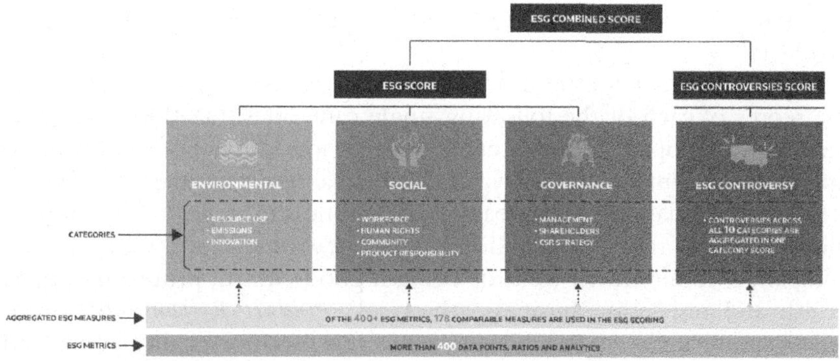

Fig. 4.5 Thomson Reuters ESG scores Methodology (*Source* Thomson Reuters Datastream)

These ESG overall scores may be mainly distinguished into two types:

I. **ESG scores:** which represent the company's ESG performance counting on reported data of public domain.
II. **ESG combined score**: which puts together the overall performance and the impact of controversies in order to offer a more comprehensive measurement of the company's sustainability conduct. Controversies are considered the events that may take place during the year such as lawsuits, legislation disputes or fines that have a negative impact on the reputation of the company.

The **first type of score** (ESG score) is grouped into ten categories where each of them includes underlying measures based on comparability, data availability and industry relevance, proportionately weighted.

More in detail, at first the Environmental factor "E" includes: (a) *Resource Use score* that reflects a company's performance and capacity to reduce the use of materials, energy or water, and to find more eco-efficient solutions by improving supply chain management; (b) *Emissions Score* represents the commitment and the effectiveness of the reduction of environmental emission in the operational processes and (c) *Innovation Score* takes account the capacity to reduce the environmental costs and burdens for its customers, thereby creating new market opportunities through new environmental technologies and processes or eco-designed products.

Again, the second letter of "S", that stays for "Social factor", considers the scores of each of the following single categories: (a) *Workforce Score*; measures a company's effectiveness towards job satisfaction, healthy and safe workplace, maintaining diversity and equal opportunities; (b) *Human Rights Score*; quantifies a company's effectiveness in respecting the fundamental human rights conventions; (c) *Community Score*; measures the company's commitment towards being a good citizen, protecting public health and respecting business ethics; (d) *Product Responsibility Score*; reflects company's capacity to guarantee and produce quality goods and services integrating the customer's health and safety, integrity and data privacy.

At last, "G" letter considers "Governance factor" that implies: (a) *Management Score*; represent a company's commitment and effectiveness towards following best practise corporate governance principles; (b) *Shareholders Score* that measures a company's effectiveness towards equal treatment of shareholders and the use of anti-takeover devices; (c) *The CSR Strategy score* that reflects a company's practises to communicate that it integrates the economic (financial), social and environmental dimensions into its day-to-day decision-making processes.

On the other hand, the ***ESG Controversies*** category is computed using 23 ESG controversy topics. This index considers the following issue: Anti-competition, business ethics, intellectual property, critical countries, public health, tax fraud, child labour, human rights, management compensation, management departures, consumer controversies, customer health and safety, privacy, product access, responsible marketing, responsible R&D, environmental, accounting, insider dealings, shareholder rights, diversity and opportunity, employee health and safety, wages and working conditions. During the year, if a scandal occurs, the company involved is penalized and this affects their overall ESG Combined score and grading. The impact of the event may last for more than a period, for example, lawsuits, ongoing legislation disputes and/or fines.

Considering the second category titled "***ESG Combined Score***", we define a score that overlays the ESG Score with ESG controversies to provide a comprehensive evaluation on the company's sustainability impact and conduct. The main objective of this score is to discount the ESG performance score based on negative media stories. It does this by incorporating the impact of significant, material ESG controversies in the overall ESG Combined Score.

Furthermore, it deserves to be mentioned another score called "**ESG Equal-Weighted Rating**", due to its relevance to empirical approaches. The ESG Equal-Weighted Rating (EWR) represents the main independent variables in the cost of equity baseline model. This score is retrieved from Thomson Reuters Asset4 database and is a robust, data-driven assessment of companies' ESG performance and capacity, where company size and transparency biases are minimal. The main differences with respect to the ESG scores are lower data-driven category weights, to reflect data availability within each category that supports more precise differentiation across companies, and the addition of Industry and Country benchmarks at the data point-scoring level, to facilitate comparable analysis within the peer group.

Sustainable Investment Strategies: Differences and Characterization

After the introduction of the different definitions of sustainable investing, together with ESG scores, the organizations promoting its diffusion and the importance of having reliable and homogenous reporting, we can surely try to better comprehend the several investment approaches that can be implemented. Although there are still some slight differences in defining the various strategies, regional and global investment forums have recently converged towards a set of seven common investment strategies, as shown in Table 4.2 (Eurosif, 2018).

The different approaches, that follows the classification introduced in 2012, can be summarized in seven categories of strategies:

- *Best-in-class*: as a positive screening, could be defined as an investment strategy centred on selecting the best performers relative to a category based on ESG metrics. It can be applied across sectors, industries or individual companies, allowing for a certain flexibility in its implementation. Often employed via linking ESG and financial metrics, it allows fund managers to include in their investable

Table 4.2 Comparing sustainable investment strategies

Eurosif	GSIA-equivalent	PRI-equivalent	EFAMA-equivalent
Exclusion of holdings from investment universe	Negative/ exclusionary screening	Negative/ exclusionary screening	Negative screening or Exclusion
Norms-based screening	Norms-based screening	Norms-based screening	Norms based approach (type of screening)
Best-in-Class investment selection	Positive/ best-in-class screening	Positive/ best-in-class screening	Best-in-Class policy (type of screening)
Sustainability themed investment	Sustainability-themed investing	Sustainability themed investing	Thematic investment (type of screening)
ESG integration	ESG integration	Integration of ESG issues	-
Engagement and voting on sustainability matters	Corporate engagement and shareholder action	Active ownership and engagement (three types): Active ownership Engagement (Proxy) voting and shareholder resolutions	Engagement (voting)
Impact investing	Impact/community investing	-	-

Source European Sustainable Investment Forum (2018) "European SRI study 2018"

universe companies operating in "less-green" sectors,[22] as long as only the most virtuous corporations are selected.

- *Sustainability themed*: investment strategy allowing investors to focus on a specific theme encompassed in the broader sustainability discussion.[23]
- *Norm-based screening*: also called international conventions, is a type of negative screening based on the adherence of a company to international norms and/or minimum standards of business practises. This kind of screening is usually implemented in combination with other strategies and commonly related to themes such as human rights, environmental protection and anti-corruption.
- *Engagement*: is a proactive and long-term oriented strategy that entails the use of shareholder activism to influence companies' stance towards sustainability themes. It can be implemented via ad-hoc discussions with the management or through more traditional means, such as filing shareholder proposals and proxy voting.
- *Exclusions*: this strategy is also called negative screening and is one of the oldest SRI strategies and can be considered as the "other side of the coin" with respect to *Best-in-class*. The main procedure consists

[22] I.e. weapons, tobacco or coal production.

[23] I.e. renewable energy, water management, energy efficiency, waste management and climate change.

in excluding from a particular fund companies, sectors or even countries that are involved in activities considered controversial or that otherwise constitute a threat from a risk management perspective. While the screening can be based on ESG or normative criteria, the most common exclusions are weaponry trade and production, tobacco, pornography, gambling, etc.

- *Impact investing*: being one the newest and yet most vibrant of the various SRI approaches, impact investing explicitly requires investors' commitment to generate tangible and measurable impact on sustainable development. Usually targeted at solving specific social or environmental problems, what differentiates impact investing from other strategies is a "dual mandate" of generating both positive financial returns and tangible ESG-measured performance improvements. In this sense, green, social and sustainability bonds are usually considered as part of impact investing.
- *ESG integration*: is the practise of explicitly and systematically integrating ESG factors in the traditional investment analysis. Given the lack of clarity in specifically defining this approach it is still quite complex to grasp the depth and diffusion of this strategy. This is because the approach can encompass everything from a basic screening method to a robust and integrated procedure throughout the investment process. Nevertheless, the general increase in interest towards ESG data, as previously highlighted, and the widespread growth in SRI-focussed functions inside investment teams, all hint towards a general positive trend in the investment industry.

The Market: Size and Characteristics

ESG factors have become an even more important part of the investment process. There are an increasing number of companies that have developed governance processes to measure, analyse, drive and communicate sustainability efforts over the last years. Not only limited to third-party analysis and the increase of public and investor demands related to ESG, corporate self-reporting is becoming more widespread among companies aiming at disclosing their contribution to sustainability development and gaining recognition from stakeholders.

Information on sustainability can serve as a differentiator in competitive industries and can foster investor's confidence, trust and employee loyalty. Hence, the benefits of non-financial reporting go beyond simply related to firm financial performance with ESG factors with the scope to maintain the licence to operate into the market.

Moreover, analysts often take account of ESG data disclosure when computing their assessment of management quality and recently, their creditworthiness. The benefits deriving from the disclosure of sustainability-related information might bring several benefits to the company issuing it. These evidences can easily translate into "value creation" features; among the others we remark the possibility to improve reputation, helping the organization to refine corporate vision and strategy, monitoring long-term risk and improves long-term risk management; leading to other forms of cost savings within the organization; helping the firms to take measures to increase long-term profitability.[24]

In the last few decades sustainable investing has grown from a niche area of the market to a relevant part of the work of both asset managers and asset owners, and finally to a larger diffusion among retail investors and the public opinion as a whole. The most recent[25] global and regional aggregate data on sustainable investing highlights that SRI strategies account for over $30 trillion,[26] a share of the total professionally managed assets (AUM) comprised between 18 and 63% based on different regions (Fig. 4.6) as well as substantial and consistent growth rates (Table 4.3).

If we consider a global scale, Europe is still the largest and most relevant market. The smallest growth rate is partly due to the maturity of the SRI market when compared with other regions but also to the common practise of applying multiple strategies to the same portfolio (GSIA, 2016) and a generally stricter definition of sustainable investing leading to a reduction of reported figures (GSIA, 2018). While its share of global SRI-managed AUM has reduced from the 52.6% of 2016 (GSIA, 2016), it is still the largest market, closely followed by the US.[27]

[24] Value of sustainability reporting (2016). A study by EY and Boston College Center for Corporate Citizenship.

[25] As of 2018.

[26] Mainly focused in Europe ($14,074 billion) followed by US ($11,995 billion) and then Japan ($2,180 billion), Canada ($1,699 billion) and Australia/New Zealand ($734 billion). Source: Global Sustainable Investment Alliance (2018) "Global Sustainable Investment Review 2018".

[27] Source: Global Sustainable Investment Alliance (2018) "Global Sustainable Investment Review 2018".

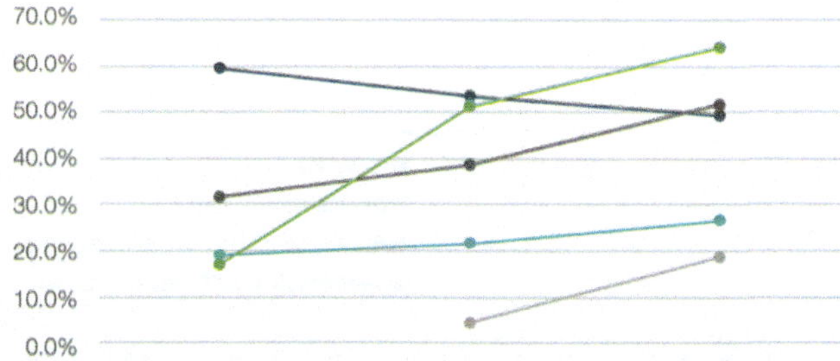

	2014	2016	2018
Europe	58.8%	52.6%	48.8%
United States	17.9%	21.6%	25.7%
Canada	31.3%	37.8%	50.6%
Australia/New Zealand	16.6%	50.6%	63.2%
Japan		3.4%	18.3%

Fig. 4.6 Percentage of sustainable investing relative to total AUM (*Source* Global Sustainable Investment Alliance [2018] "Global Sustainable Investment Review 2018")

Table 4.3 Growth of sustainable investing 2014–2018 (data in billion)

	2014	2016	2018	Growth Per Period: Growth 2014–2016	Growth Per Period: Growth 2016–2018	Compound Annual Growth Rate (CAGR) 2014–2018
Europe	€ 9,885	€ 11,045	€ 12,306	12%	11%	6%
United States	$ 6,572	$ 8,723	$ 11,995	33%	38%	16%
Canada (in CAD)	$ 1,011	$ 1,505	$ 2,132	49%	42%	21%
Australia/New Zealand (in AUD)	$ 203	$ 707	$ 1,033	248%	46%	50%
Japan	¥ 840	¥57,056	¥231,952	6692%	307%	308%

Source Global Sustainable Investment Alliance (2018) "Global Sustainable Investment Review 2018"

As we can note, negative screening is by far the most relevant in terms of assets, closely followed by ESG integration (Fig. 4.7) that, instead, registered a CAGR[28] of 27% during the last three years (Fig. 4.8). Among the most widely adopted strategies, ESG integration has had the largest growth rate from 2016.[29] With regards to the importance of each

[28] Compound Annual Growth Rate.

[29] Registering a growth rate 2016–2018 equal to 69%.

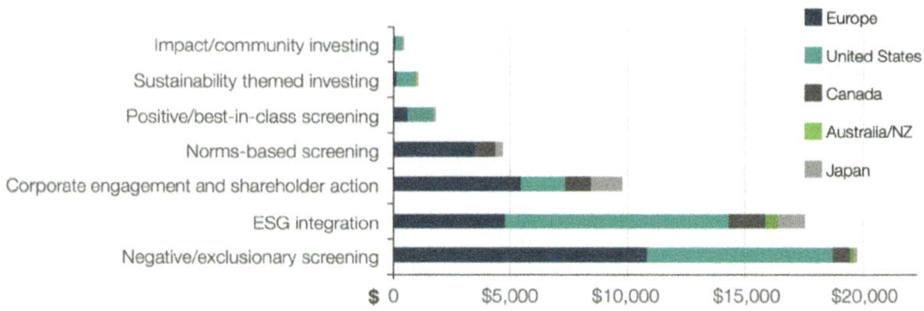

Fig. 4.7 SRI assets by strategy and region (data in billion) (*Source* Global Sustainable Investment Alliance [2018] "Global Sustainable Investment Review 2018")

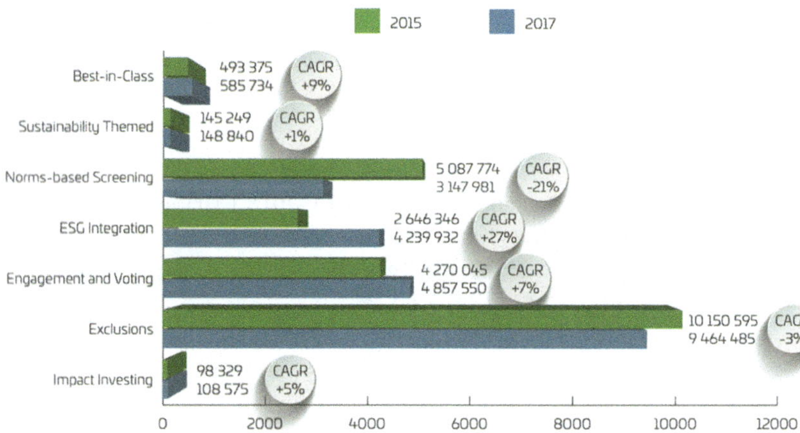

Fig. 4.8 European SRI AUM breakdown by strategy (data in € million) (*Source* European Sustainable Investment Forum [2018] "European SRI study 2018")

strategy for the various global regions, while Europe leads in both engagement and exclusion, the US holds the largest proportion of assets in the remaining strategies.

References

Baumann-Pauly, D., Wickert, C., Spence, L. J., & Scherer, A. G. (2013). Organizing corporate social responsibility in small and large firms: Size matters. *Journal of business ethics 115*(4): 693–705.

Black, F., & Scholes, M. (1973). The pricing of options and corporate liabilities. *Journal of Political Economy, 81*(3), 637–654.

Blowfield, M., & Murray, A. (2014). *Corporate responsibility*. Oxford University Press.

Bowen, H. R. (1953). *Social responsibilities of the businessman*. Harper & Row.

BrownFlynn. (2018). The ESG ecosystem understanding the dynamics of the sustainability ratings & rankings landscape February 2018.

Carroll, A. B. (1999). Corporate social responsibility: Evolution of a definitional construct. *Business and Society, 38*(3), 268–295.

CFA Institute. (2015). *Environmental, social, and governance issues in investing: A guide for investment professionals*. https://www.cfainstitute.org/-/media/documents/article/position-paper/esg-issues-in-investing-a-guide-for-investment-professionals.ashx

Chatterji, A. K., Durand, R., Levine, D. I., & Touboul, S. (2016). Do ratings of firms converge? Implications for managers, investors and strategy researchers. *Strategic Management Journal, 37*(8), 1597–1614.

Cheng, B., Ioannou, I., & Serafeim, G. (2014). Corporate social responsibility and access to finance. *Strategic Management Journal, 35*(1), 1–23.

Davis, K. (1973). The case for and against business assumption of social responsibilities. *Academy of Management Journal, 16*(2), 312–322.

Directive 2014/95/EU. (2017). *A comprehensive overview of how member states are implementing the EU directive on non-financial and diversity information.*

Eccles, R. G., & Stroehle, J. C. (2018). *Exploring social origins in the construction of ESG measures*. Available at SSRN 3212685.

Elkington, J. (1997). *Cannibals with forks. The triple bottom line of 21st century*, 73.

European Sustainable Investment Forum. (2018). *European SRI Study 2018*. http://www.eurosif.org/wp-content/uploads/2018/11/European-SRI-2018-Study.pdf

EU Directive disclosure of non-financial information and diversity information—EY (2017).

Fassin, Y. (2008). SMEs and the fallacy of formalising CSR. *Business Ethics: A European Review, 17*(4), 364–378.

Freeman, R. E. (1984). *Strategic management: A stakeholder approach*. Pitman.

Friede, G., Busch, T., & Bassen, A. (2015). ESG and financial performance: Aggregated evidence from more than 2000 empirical studies. *Journal of Sustainable Finance & Investment, 5*(4), 210–233.

Friedman, M. (1970). A Friedman doctrine: The social responsibility of business is to increase its profits. *The New York times Magazine, 13*, 32–33.
Georg Kell. (2018). The remarkable rise of ESG investing. *Forbes*, 11 July 2018.
Global Sustainable Investment Alliance. (2016). *Global sustainable investment review 2016*. http://www.gsi-alliance.org/wp-content/uploads/2017/03/GSIR_Review2016.F.pdf
Global Sustainable Investment Alliance. (2018). *Global sustainable investment review 2018*. http://www.gsi-alliance.org/wp-content/uploads/2019/06/GSIR_Review2018F.pdf
Ho, V. H. (2017). Comply or explain and the future of nonfinancial reporting. *Lewis & Clark Law Review, 21*, 317.
iShare. (2019). An evolution in ESG Indexing. Available at https://www.ishares.com/us/literature/
Jeffwitz, C., & Gregor, F. (2017). *Comparing the implementation of the EU non-financial reporting directive*. Available at SSRN 3083368.
Jensen, M. C., & Meckling, W. H. (1976). Theory of the firm: Managerial behavior, agency costs and ownership structure. *Journal of Financial Economics, 3*(4), 305–360.
Lintner, J. (1965). Security prices, risk, and maximal gains from diversification. *The Journal of Finance, 20*(4), 587–615.
Llewellyn, J. (2007). *The business of climate change: challenges and opportunities*.
Markowitz, H. (1952). The utility of wealth. *Journal of Political Economy, 60*(2), 151–158.
Merton, R. (1973). Theory of rational option pricing. *Bell Journal of Economics and Management Science, 4*(1), 141–183.
Modigliani, F., & Miller, M. H. (1958). The cost of capital, corporation finance and the theory of investment. *The American Economic Review, 48*(3), 261–297.
Modigliani, F., & Miller, M. H. (1963). Corporate income taxes and the cost of capital: A correction. *The American Economic Review, 53*(3), 433–443.
Mossin, J. (1966). Equilibrium in a capital asset market. *Econometrica: Journal of the Econometric Society 34*(4), 768–783.
Principles for Responsible Investment. (2020). *PRI brochure*. https://www.unpri.org/download?ac=6303
Sharpe, W. F. (1964). Capital asset prices: A theory of market equilibrium under conditions of risk. *The Journal of Finance, 19*(3), 425–442.
Sheehy, B. (2015). Defining CSR: Problems and solutions. *Journal of Business Ethics, 131*(3), 625–648.
Task Force on Climate-Related Financial Disclosures. (2020). *About the task force*. https://www.fsb-tcfd.org/about/#

Wickert, C., Scherer, A. G., & Spence, L. J. (2016). Walking and talking corporate social responsibility: Implications of firm size and organizational cost. *Journal of Management Studies, 53*(7), 1169–1196.

Zingales, L. (2000). In search of new foundations. *The Journal of Finance, 55*(4), 1623–1653.

http://ec.europa.eu/growth/industry/corporate-social-responsibility/.

https://www.ga-institute.com/press-releases/article/flash-report-85-of-sp-500-indexR-companies-publish-sustainability-reports-in-2017.html.

https://www.unglobalcompact.org/.

https://www.globalreporting.org/.

https://ec.europa.eu/info/index_en.

CHAPTER 5

Consortium Company, Circular Economy and ESG: The Comieco Case Study

Carlo Montalbetti

Is Sustainability More a Story Than a Reality?

In the absence of an internationally recognized definition of the ESG concept, is it possible to rely on ESG factors to achieve an ethical rating and high performance? A starting point for making ESG investments a tangible reality is to verify whether the Circular Economy is a successfully applied economic, environmental and social concept and not just a policy agenda. In this context the Italian case of Comieco can be an interesting issue for thought. Comieco is the Consortium of the paper supply chain for the recycling of paper and cardboard and is part of the Conai system—the National Packaging Consortium.

C. Montalbetti (✉)
Research Centre of Applied Economics, Catholic University of the Sacred Heart, Milan, Italy
e-mail: montalbetti@comieco.org

© The Author(s), under exclusive license to Springer Nature Switzerland AG 2022

C. B. Pellegrini et al. (eds.), *Climate Change Adaptation, Governance and New Issues of Value*, Palgrave Studies in Impact Finance,
https://doi.org/10.1007/978-3-030-90115-8_5

Looking at the Last Century

In 1972 we can find the first strong indicators of the environmental risks caused by economic development due to indiscriminate consumption of natural resources: these are the conclusions of the report presented by a group of scholars appointed by the Club of Rome.[1] This report based on the analysis of different scenarios reflecting most important global trends introduces the concept of sustainable development as a strategic approach to safeguarding civil well-being and the planet.

Thanks to the report of the Club of Rome, for the first time, the model based on a linear and continuous growth of economy and population has been questioned and the availability of natural resources has been defined as one of the main limits which humanity has to face.[2]

In the 1980s, considering the countries with the highest development index, including Italy, it is evident how the exponential growth of waste (with an average annual increase of over 5%) resulting from the mass culture of "disposable" is causing significant environmental and social damage.

However, the sensitivity to these issues remained confined to restricted circles of scholars and the first environmental associations, especially in Italy. In addition, the regulations on environmental policies were weak and focused only on waste disposal and did not consider the recycling of raw materials.

In this context of a lack of interest, it is interesting to note that 12 entrepreneurs in the paper packaging sector decided to focus on the sustainable characteristics of paper and cardboard. The paper is a raw material derived from cellulose, biodegradable and recyclable: an alternative to plastics of fossil origin, whose indiscriminate and massive use (world production went from 60 to 105 million tons between 1980 and 1990) was beginning to cause alarm.

But there is not only this intuition; alongside the awareness of the environmental value of the raw material, this group of entrepreneurs denounces the use of landfills, which causes the destruction of millions

[1] The Club of Rome is a non-governmental, non-profit association of scientists, economists, businessmen, civil rights activists and heads of state from all five continents. The Club's founders were Aurelio Peccei, Alexander King and Daniel Rockefeller.

[2] The limits to Growth, known as the Meadows Report, 1972.

of tons of raw materials, such as paper and cardboard, that is essential for industrial recycling. Indeed, the industrial use of waste paper has a strategic value for the Italian paper industry because Italy needs a constant supply of recovered paper and cardboard since it is poor of forests producing large amounts of cellulose-rich biomass.[3] Moreover, in the past Italy was obligated to import from abroad as it did not have an internal collection system.

For the reason mentioned above, Comieco (Comitato per l'Imballo Ecologico) a Committee for ecological packaging was founded in April 1985, in Milan.

Comieco represents the first European group of entrepreneurs with the aim of answering the issue of the use of natural resources and stating the urgency of public policies in order to activate separate collections and a Circular Economy (CE) of cellulosic packaging.[4]

The entrepreneurial intuition of the paper industry was accompanied in the following years by a growing environmental awareness, well represented by the diffusion and establishment of environmental associations. The Italian regulatory framework was also beginning to change through the regulation governing the disposal of industrial waste and waste oils and the establishment of the obligatory national Consortium Replastic.[5]

At the same time, at the international level, for the first-time environmental policies are no longer addressed only from the point of view of "conservation", but become part of the broader concept of sustainability which also includes economic and social development.

In this context, the European Parliament, with the Italian rapporteur Luigi Vertemati, approved the packaging directive in 1994[6] introducing the principle of "Extended producer responsibility" (EPR) through material recovery and recycling and prevention as a reduction in the use of raw materials and the product eco-design.

The European business system was beginning to internalize the need to address the issue of environmental sustainability.

Despite steps forward, the political and cultural debate is still far from bringing concrete results. First of all, waste management in Italy is still

[3] Italy does not extract cellulose.

[4] Quel che resta di un bene by Carlo Montalbetti and Ercole Sori, Il Mulino, 2011.

[5] Decree-Law No. 307/1988, Legislative Decree No. 95/1992.

[6] Directive 94/62 / EC on packaging and packaging waste.

"primitive": in 1997 about 80% of the 26.6 million tons of municipal waste was still disposed of in landfills, 7% was incinerated in 38 plants, of which only 23 with energy recovery and only 9.4% was sent to recycling and composting plants. As for the paper sector, more than 4.4 million tons of cellulosic material were destroyed in landfills.

The Ronchi Decree, Conai and the Fundamentals of the Circular Economy in the Packaging Sector

The strong need to implement the European Directive on packaging has committed the Prodi government, through its Minister of the Environment Edo Ronchi, to launch a decree[7] which represents the first attempt to rationalize the entire discipline, introducing the rule of the three Rs: Reuse, Recovery and Recycling, as an alternative to landfill and incineration.

The political intuition of the measure, which will prove to be successful, is the delegation to the business system to organize the management according to a consortium approach in order to comply with the producer's responsibility keeping the control function in the hands of the public authority.

The Conai system was created following the above-mentioned Ronchi Decree, and it is a system composed of companies that produce and use packaging. The Conai system is divided into seven material consortia[8] and it is financed by contributions paid by member companies in order to support the activities of separate collection and recycling of packaging waste (Table 5.1).

Conai's main task is to ensure the achievement of national recovery and recycling targets by directing the activity of the consortia and using the framework agreement with Anci—the national association of Italian municipalities to support and develop the separate collection system.

[7] Legislative Decree 22/1997.

[8] Cial (aluminum), Comieco (paper), Coreve (glass), Ricrea (steel), Corepla (plastic) and Rilegno (wood). To these six initial consortia, in 2020, with the decree of the Minister of the Environment and the Protection of the Territory and the Sea of October 16, 2020, is added Biorepack (biodegradable and compostable plastic), the new supply chain consortium for the management at the end life of biodegradable and compostable plastic packaging collected with the organic fraction of waste.

Table 5.1 The Conai Environmental Contribution 2021

Materials	*€/t*
Steel	18.00
Aluminum	15.00
Paper	Excluding CPL[a]: 25.00 CPL: 45.00
Wood	9.00
Plastic	Level A: 150.00 Level B1: 208.00 Level B2: 560.00 Level C: 660.00
Biodegradable and compostable plastic	294
Glass	37.00

[a]CLP: containers for liquids
Source CONAI, English version

The take-off of the Conai system based on consortia was the starting point for the realization of an actual process of national CE focused on: (1) the widespread organization of separate collections, financed by consortium systems through the CAC (Conai Environmental Contribution); (2) the development of essential systems for the treatment of material flows from collections; (3) the consolidation of the national recycling industry and (4) the articulation of prevention as a tool for the sustainable redesign of packaging products.

The Comieco Case: The Circular Economy Is Not a Mirage but an Opportunity

The progress of the Conai system has led to the transition of the Committee for Ecological Packaging into the National Consortium for the recovery and recycling of cellulose-based packaging: Comieco has expanded from a small group of companies into a Consortium with over 3,000 companies. Therefore, Comieco represents the paper packaging supply chain and is responsible for achieving public goals. In addition, it supports local authorities providing them the necessary guarantees of collection of the material collected, with recognition of a fee for the service provided.

Comieco has transformed its mission over time, if previously it offered an exclusive service only to its consortium members, today it plays a role of public interest keeping the legal personality of private law. Specifically, Comieco is committed to: (a) environmental protection (in compliance with what is prescribed by the EU); (b) the economic operation of the whole cellulosic-based packaging chain (in which all economic operators involved are interested) and (c) the competitiveness of the consortium companies (which are highly exposed to market competition).[9]

Thanks to consortium system, many technical, organizational and managerial innovations have been introduced in Italy that have allowed the supply chain to embark on a path towards the CE capable of considering and governing the interrelationships between environmental, social and economic components.

First of all, the development of an integrated and coordinated waste collection and management system was accompanied and supported, ranging from the collection of waste to its treatment, recycling and economic valorization. All this has been integrated with a strong communication and awareness activity towards the final consumer in order to involve him in the separate collection process.

Comieco has integrated private core with a public core generating a model characterized by strong synergies. The private components pushed the Consortium to operate efficiently and effectively in decision-making and operations, whereas the public one contributed to defining a mission focused on environment protection and conservation of natural resources. The latter component has allowed the development of a circular model that can guarantee—under the same conditions throughout the country—the collection and effective recycling of cellulosic-based packaging, going beyond the objectives set by the legislation.

The operation of the system has therefore made it possible to achieve three strategic results: (1) drastically reduce the disposal of precious second raw materials in landfills; (2) develop a separate collection of paper and cardboard per capita from 17 kg per inhabitant in 1998 to 57.2 kg per inhabitant in 2020 and (3) to guarantee the Italian paper industry, since 2005, the supply of waste paper from separate collection without depending on imports from abroad (Table 5.2).

This complex and important activity is carried out according to the subsidiarity principle. In the paper and cardboard supply chain the demand for waste paper is represented by paper mills, whereas the

[9] Vittorio Coda, Imprenditorialità consortile ed economia circolare nella filiera cartaria. Il caso Comieco, edited by Mario Minoja and Giulia Romano, Egea, 2020.

Table 5.2 Imports and exports of waste paper

Import–export of waste paper (Italy, 2013–2020)		
Year	*Import (ton)*	*Export (ton)*
2013	338,000	1,685,000
2014	310,000	1,678,000
2015	325,000	1,822,000
2016	384,000	1,940,000
2017	355,000	1,867,000
2018	401,000	1,903,000
2019	311,000	1,815,000
2020	247,000	1,812,000

Source Comieco

supply comes from recyclers and municipal separate collection. Therefore, Comieco can be considered as an intermediary between supply and demand remaining subsidiary to the market: packaging producers comply with the obligations provided by legislation on recovery and recycling ensuring a remuneration to municipalities for the collection activities carried out through the Conai environmental contribution.

If the demand and supply of waste paper find by themselves a point of equilibrium that guarantees recycling in line with regulatory obligations, so Comieco "steps aside".

Otherwise, Comieco buy the waste paper at conditions that allow the municipalities to cover the higher costs of separate collection (with the latest agreement signed, at least 80% of the costs) and paper mills to find convenient its use as a secondary raw material.

The public utility role played by Comieco has gone beyond its strictly technical ecological "mission". In fact, it has not only allowed to divert cellulose waste from landfills but also to create a stable supply of waste paper for the Italian paper industry increasing its competitiveness and, to improve Italy's balance of payments by transforming it from an importer to a net exporter of waste paper, also thanks to the progressive improvement of its quality.

The consortium system was crucial in 2009–2010, when the financial crisis, which later became the global economic crisis, led to a sharp reduction in packaging demand.

Comieco has made it possible to overcome the crisis phase without blocking separate collections in Italian municipalities, through the obligation imposed on the consortium Italian paper mills to guarantee a recovery and recycling quota in accordance with the principle of extended producer responsibility.

Moreover, the consortium's public mission requires it to cover the entire country homogeneously, avoiding the phenomenon of "cherry-picking" choosing to operate only in the most profitable areas or where it is easier to do so.

The collection of the waste paper throughout the country under the same conditions is also a guaranteed item for the environment, because it allows to develop separate collection everywhere, guaranteeing municipalities the possibility of delivering cellulosic waste at pre-established conditions that cover most of the costs of the collection itself.

The universal service obligation has characterized the consortium system for packaging management from the outset. Comieco has not met this obligation passively, waiting for the requests from the municipalities of Southern Italy and the Islands, but has acted autonomously to promote separate collection in these areas with investments of around 10 million euros, allowing a substantial development of the southern areas which in the last three years have seen a net growth rate of almost 30%.

The consortium system, as it did at the peak of the economic crisis in 2009–2010, has also shown resilience during the recent lockdown period due to the Covid-19 pandemic in the spring of 2020.

Comieco has incentivized paper mills to collect the waste paper (also selling it at negative prices) and pay the services to the recovery plants that transform the waste into waste paper.

Thanks to the use of budgetary reserves and the increase in the Conai environmental contribution, the consortium continued to collect cellulose waste and to pay the higher costs for separate collection to municipalities, avoiding disposal in landfills or incineration, given that the value of waste paper was below zero in Italy and Europe.

In the highlight of the facts, Comieco can be considered a successful example of a "consortium company".[10] It can be seen that some important elements of the ESG principles are already operative within the Comieco by looking the Specific Prevention Plan (PSP). The latter is a legal planning and reporting tool that the Consortia, through Conai, submit to the Ministries of Ecological Transition and the Ministry of Economic Development.

[10] Imprenditorialità consortile ed economia circolare nella filiera cartaria, il caso Comieco by Mario Minoja and Giulia Romano, Egea Editions, 2020.

The PSP indicates both the recycling targets achieved and the related economic commitments made, as well as all the activities carried out to maintain active and collaborative relationships with stakeholders.[11]

In fact, the CE generated by cellulosic waste as well as by other packaging materials can work thanks to the active involvement of public and private entities which are part of the paper and cardboard supply chain. Therefore, it is possible to define it as the "Comieco system", a complex network of entities, roles and relationships involving the entire supply chain and its stakeholders at every stage of processing.

Specifically, the key players are:

- final consumers, citizens and companies that separate the paper and cardboard used at source;
- municipalities and managers of the service for the collection and disposal of urban and assimilated waste, which organize the separate collection of paper and cardboard, often combining graphic paper with packaging;
- waste recovery plants, the so-called "platforms" (they can be either private companies or plants owned by local entities), which select and transform paper and cardboard from separate collection into waste paper according to instructions of paper mills;
- paper mills, which purchase waste paper to produce new paper in reels;
- participants in the auctions of the collected waste paper, which can be paper mills, waste management plants and brokers who act as intermediaries between those who produce the cellulosic waste and those who want to buy it for recycling (paper mills/plants);
- converters (or paper converting companies), i.e., packaging producers who pay the environmental contribution requested by Conai to Comieco; and
- user companies, which purchase cellulose-based packaging from converters to package their products.

[11] Community and territory.

In terms of governance, the Statute (approved by the Ministries of Ecological Transition and the Ministry of Economic Development) recognizes the equal representation of stakeholders in the Comieco model regardless of the number of consortium members in each category.

Comieco is a consortium that manages a complex supply chain and for this reason conflicts of interest may arise between consortium members and affiliates, and so must be managed.

However, these conflicts of interest can represent an opportunity because they allow to understand how certain choices and decisions impact on the entire system and several activities of the consortium.

In compliance with the Consortium's public utility goals, the reporting systems used have always considered environmental and social factors as required by CSR and ESG systems.

From the Circular Economy to the Green Deal

Italy has achieved important results in a rather short time; thanks to the consortium system that supports the municipalities in the management of separate collection and guarantees a value through recycling. In the last 20 years, per capita separate collection of paper and cardboard has continuously grown, from 17 kg per inhabitant year in 1998 to 57.2 kg per inhabitant year in 2020. Today our country is a net exporter of waste paper for recycling.

However, today the European reference framework is considerably raising the challenge in terms of quality and prevention in order to increase the circularity of processes.

A first effort towards a CE by the European Commission can be identified with Communication 398/2014 "Towards a circular economy: A zero waste program for Europe", defining the basis for the subsequent regulatory framework on CE approved after a long and complex process in December 2015.

Four directives belonging to the "Circular economy package" came into force on 4 July 2018. In particular, the directive on packaging (Directive 852/2018) sets the general recycling target of 65% by 2025 and 70% by 2030, identifying specific targets for different materials.

The paper supply chain has already reached and exceeded the 75% recycling target set for paper and cardboard by 2025 and being 81% for recycling, it is in line with the 85% target set for 2030.

The CE represents a new economic paradigm of reference for the development of sustainable business models and models of capitalism.

The circular business model considers waste as a potential resource, provided it is collected, treated and recycled. Waste becomes "second raw material" able to generate economic value and create new possibilities of economic, environmental and social development.

In Italy the CE is well consolidated and is equipped with a strong recycling industry[12] with good growth opportunities in order to become the leader of Southern Europe.

The Green New Deal and the Green Transition, funded with 37% of European funds over the next years, identify waste prevention and treatment, CE and bio-economy as key field of action to mitigate climate change.

These fields of action correspond to the goals defined by both the European Union and the Italian Government within the framework of the "National Recovery and Resilience Plans" and the Recovery Fund in order to relaunch the economy in Europe.

This is a great opportunity for the paper and board supply chain to improve its potential for innovation and development in the CE.

The first challenge involves the evolution of the regulatory framework on recycling, promoting the recycling of cellulose-based products such as graphic paper and paper for sanitary use, and not only the recycling of packaging.

The second challenge is to improve the quality and "purity" of packaging waste and cellulosic materials in general to increase their recyclability.

A third challenge is the creation of new supply chains of bio-based, recycled and recyclable packaging supported by research and development of new materials that are more eco-compatible and so recyclable, of packaging with lower weights, of packaging that helps citizens in the right separation of waste at source.

These challenges imply a business model with an ESG approach. In the context of Comieco, this means enhancing its strategic assets, first of all there is the ecological mission. Moreover, for Comieco it is important to promote the competitiveness of the paper and cardboard supply chain,

[12] Economia circolare in Italia by Duccio Bianchi, Edizioni Ambiente, 2019.

to improve Italy's trade balance and to spread civic awareness, demonstrating that the market and public interest service can work together in the pursuit of common benefit.

References

Bianchi, D. (2019). *Economia circolare in Italia. La filiera del riciclo asse portante di un'economia senza rifiuti*. Edizioni Ambiente.

Decree-Law No. 307/1988—Legislative Decree No. 95/1992.

Directive 94/62 / EC on packaging and packaging waste.

Meadows, D. H., Randers, J., & Meadows, D. L. (1972). *The limits to growth—A report for the club of Rome's project on the predicament of Mankind*. New York: Universe Books. ISBN 0876631650.

Minoja, M., & Romano, G. (2020). *Imprenditorialità consortile ed economia circolare nella filiera cartaria*. Il caso Comieco. Egea.

Montalbetti, C., & Sori, E. (2011). *Quel che resta di un bene*. Il Mulino.

CHAPTER 6

Environmental, Social, and Governance Issues: An Empirical Literature Review Around the World

Claudia Cannas, Maurizio Dallocchio, and Laura Pellegrini

Introduction

The economic literature on Environmental, Social, and Governance (ESG) issues has seen fast growth in recent decades. In this context, particular attention was paid to the role of the company and the impact of its strategic choices.

C. Cannas wrote the following sections: A New Paradigm of Company, ESG Factors and Cost of Capital, Sustainability and Stock Prices.
M. Dallocchio wrote the following sections: Introduction, Conclusions.
L. Pellegrini wrote the following sections: ESG Factors and Firms' Performance.

C. Cannas (✉)
Research Centre of Applied Economics, Catholic University of the Sacred Heart, Milan, Italy
e-mail: claudia.cannas@unicatt.it

© The Author(s), under exclusive license to Springer Nature Switzerland AG 2022

C. B. Pellegrini et al. (eds.), *Climate Change Adaptation, Governance and New Issues of Value*, Palgrave Studies in Impact Finance,
https://doi.org/10.1007/978-3-030-90115-8_6

These studies embrace the concept of Corporate Social Responsibility (CSR) which was developed over time in several ways. Over time, many authors attempted to explain their concept of CSR and the important role of companies. Among the numerous definitions of Corporate Social Responsibility (CSR) we proposed that of Davis, one of the first and most prominent scholars during the 1960s:

> It is the firm's obligation to evaluate in its decision making process the effects of its decisions on the external social system in a manner that will accomplish social benefits along with the traditional economic gains which the firm seeks. It means that social responsibility begins where the law ends. A firm is not being socially responsible if it merely complies with the minimum requirements of the law, because this is what any good citizen would do. Social responsibility goes one step further. It is a firm's acceptance of a social obligation beyond the requirements of the law.

According to Davis (1973), Corporate Social Responsibility should be interpreted in a managerial context and it can be justified only in the light of a long-run economic gain to the firm.

However, there were contrary opinions about the issue. Corporate Governance considers two main theories which have guided companies' business strategy during the time, as well as the *Shareholder Theory* (Friedman, 1970) and the *Stakeholder Theory* (Freeman, 1984). In the well-known article in Financial Times in 1970, Milton Friedman affirmed that only people have responsibilities, not companies.

He stated:

> There is one and only one social responsibility of business—to use its resources and engage in activities designed to increase its profits so long as

---

M. Dallocchio
Department of Finance, Bocconi University, Milan, Italy
e-mail: maurizio.dallocchio@sdabocconi.it

L. Pellegrini
Department of Economics and Finance, University of Bari "Aldo Moro", Bari, Italy
e-mail: laura.pellegrini@uniba.it

it stays within the rules of the game, which is to say, engages in open and free competition without deception or fraud.

This perspective, known in literature as *Shareholder Theory*, argues that Firms have the only responsibility to generate value for their Shareholders.

On the other hand, the *Stakeholder theory* going against this typical neoclassical view of profit maximization and shareholder value as the main purpose of the company has prompted the literature to investigate the impact of corporate choice on all stakeholders. In particular, Freeman extended the role of the company beyond the economic dimension considering the multiplicity of individuals with which the company relates but also the relationship between the company and the socio-economic environment. Based on the development of the Corporate Social Responsibility concept, both businesses and institutions started a process towards the integration of ESG issues into their decision-making process.

The awareness that we live in a world where various resources have started to be in short supply due to not only the limited availability of natural resources but also human actions have determined significant unsustainable conditions on the planet.

In this difficult context, both scholars and practitioners have tried to respond to the problem of the sustainability of the world's systems and economies by attempting to indicate a possible way forward. For almost a century, however, corporate social responsibility and doing business by respecting ethical principles have become a public concern.

In this chapter, we have tried to parsimoniously collect the main studies on ESG issues, while the first strand of literature focussed on the effect that ESG scores have on the cost of capital (Bauer & Hann, 2010; Bhojraj & Sengupta, 2003; Schauten & Van Dijk, 2011) and whether this information increases the benefits for equity investors (Bebchuk et al., 2009; Cremers & Ferrell, 2014; Gompers et al., 2003). Then, a more recent literature analysed the relationship between the development of ESG practices and Firms' Operating Performance with the aim to verify whether greater ESG practices are capable to determine better performances creating long-term value (Fulton et al., 2012; Margolis et al., 2007; Van Beurden & Gössling, 2008).

A New Paradigm of Company

The idea of new paradigm of company capable to embrace the three-dimensional concept of sustainability (Elkington, 1994–1997) has prompted academic literature to analyse the impact of companies' activities at environmental, social, and governance level. The concept of acting in a responsible way has taken different patterns through the years. From the beginning of the new millennia, greater attention to the ESG issues and sustainability has been paid by mainstream institutions.

Therefore, even if all studies don't consider at the same time all three dimensions of sustainability, it seems possible to observe that generally companies with higher sustainability scores achieve better performance.

First studies in this field of research investigate on the ESG scores impact on the cost of capital such as cost of equity and cost of debt and on the related risk. Empirical evidences highlight that companies that adopt greater sustainability practices and policies significantly reduce both cost of debt and cost of equity (Bauer & Hann, 2010; Bhojraj & Sengupta, 2003; Schauten & Van Dijk, 2011). In particular, the adoption of better sustainability standards in terms of a good corporate governance structure (Bhojraj & Sengupta, 2003) and disclosure policies (Soppe et al., 2011) reduces the cost of borrowing.

ESG Factors and Cost of Capital

There is an extensive literature that analyses how environmental issues impact Firm Performance captured cost of capital. In particular, with regard to the environmental dimension, Konar S. and Cohen M. A. (2001) observe that environmental performance has a significantly negative impact both economically and statistically on the value of intangible assets traded by companies included in the S&P500 index. Specifically, while the effect of environmental controversies on the value of intangibles is statistically significant but not economically significant in a large number of industries, the effect of pollutant emissions, on the other hand, tends to be both statistically and economically significant.

Evidences seem to suggest that larger firms are rewarded for voluntary and increased compliance towards environmental issues, both in terms of financial and reputational performance. In 1972, Bragdon and Marlin found that there was a significant relationship between environmental performance and firm financial performance. However, Chen and

Metcalf's (1980) analysis based on the same sample shows that this relationship is not significant when firm size is taken into account.

Moreover, several studies point out also that the development of a good corporate governance leads to lower cost of equity (Lima & Sanvincente, 2013), environmental risk management practices, disclosure on environmental policies (El Ghoul et al., 2018), good employee relations, and product safety (El Ghoul et al., 2011) lower firm's cost of equity.

Barth et al. (2013) used earnings transparency measure as an indicator of quality for governance providing evidence that firms more transparency enjoy a lower cost of capital. Dhaliwal et al. (2011) examine whether there is a reduction in firm's cost of equity capital associated with initiation of voluntary disclosure of corporate social responsibility activities. They argue that CSR practises affect Firms' Performance and Valuation because such activities can help firms avoid potential government regulation and reduce future compliance costs.

Based on the assumption that voluntary disclosure reduces information asymmetry, Dhaliwal et al. (2011) used a large sample of firms rated by KLD Research and Analytics Inc. to regress the voluntary disclosure measured as a dummy variable (taking the value of one when a firm first issues a standalone CSR report) against the corporate social performance. Through their findings they show that firms with high cost of equity in the previous year tend to initiate disclosure of CSR activities in the current year and that among the initiating firms the ones characterized by superior social responsibility performance enjoy an additional reduction in the cost of equity.

Other evidences concerning this issue are provided by Reverte (2012), who examines how the quality of CSR reporting impacts on the cost of equity for a sample of Spanish-listed firms. These findings contribute to the debate on whether CSR activities contribute to add value or they are value-neutral showing that enhancing CSR can increase firm value by reducing the firm's cost of equity.

Finally, El Ghoul et al. (2011) examine the impact of CSR on the cost of equity for a large sample of US firms. Through different approaches to estimate firms' ex-ante cost of equity, they point out that firms with better CSR ranking show cheaper equity financing. Specifically, they highlight that investment in improving responsible employee relations, environmental policies, and product strategies reduce Firms' cost of equity. On the contrary, firms operating in tobacco and nuclear power industries increase firms' cost of equity by supporting in this way the arguments that firms with socially responsible practices have higher valuation and lower risk.

Sustainability and Stock Prices

Further studies have attempted to identify the benefits for investors related to implementation of ESG factors into investment choices. However, the general perception of investors seems to deviate from the idea of a positive impact of these factors.

The stock market is highly sensitive to news available in the market and the behaviour of economic actors. The prices incorporate news and by using methodological approaches such as event studies or portfolio analyses it is possible to analyse the effect of new information deriving from a specific event in a specific stock. These approaches have been widely used to analyse environmentally or socially related events.

Aktas et al. (2011) argue that mergers and acquisitions offer a framework on the stock market performance of socially responsible investments (SRI). Findings suggest that the stock market rewards the acquirer for making socially and environmentally responsible investments. They also document that environmental and social performance of the acquirer increases following the acquisition of an SRI-aware target.

Bebchuk et al. (2009) stated that financial markets assess better governance quality positively. They consider an index based on two groups of provisions: the first entails four constitutional provisions that limit the careless behaviour of shareholders (staggered boards, limits to shareholders bylaw amendments, supermajority requirements for mergers, and supermajority requirements for charter amendments), while the second group includes two measures anti-takeover (poison pills and golden parachutes). They argue that increases in the level of this index are monotonically associated with a significant reduction in Firm Valuation captured by Tobin's Q. In their findings, Bebchuk et al. (2009) find that firms with higher levels of the entrenchment index are associated with large negative abnormal returns. In addition, by examining all sub-periods, they provide evidence that a strategy of buying low entrenchment firms and selling short high entrenchment firms out-performs the market in most such periods and does not underperform the market in even a single sub-period.

Brammer et al. (2006) investigate the relationship between Corporate Social Performance and Financial Performance using stock returns as a measure, for a sample of UK-listed companies. Their main findings underline that firms with higher social performance scores tend to achieve lower returns, while firms with the lowest CSP scores notably outperformed the market. Therefore, they document that the environmental and employment indicators are negatively related with returns while the community indicator is weakly positively correlated.

We highlight that also a negative reaction from the market has taken place in cases of voluntary adoption of corporate environmental initiatives. Fisher-Vanden and Thorburn (2011) considering a sample of 117 firms over 1993–2008 observe that shareholder wealth effects result from participation in the voluntary environmental activities by using an event study methodology. Specifically, they argue that companies that announce membership in EPA's Climate Leaders[1] experience significantly negative abnormal stock returns and that the price decline is larger in firms with poor corporate governance structures and for high market-to-book firms. Overall, corporate commitments to reduce greenhouse gas emissions appear to conflict with firm value maximization.

ESG Factors and Firms' Performance

This biased investor view is primarily due to evidence from a number of portfolio studies showing a neutral relationship in terms of the impact of ESG on Firm Performance.

Waddock and Graves (1997) argue that ESG-related costs reduce profits and shareholders' wealth. Campbell (2007) also assumes that ESG activities are similar to a form of corporate charity. However, although there are several studies that suggest a negative relationship or results that are not statistically significant (Horvathova, 2010), there are more than 2,100 other empirical studies (see Margolis et al., 2007) – particularly company-based – that instead suggest a positive ESG relationship (Friede et al., 2015), albeit with a moderate and possible decrease over time (see also Orlitzky et al., 2003).

[1] The Environmental Protection Agency (EPA) of Center for Corporate Climate Leadership is a resource center for all organizations looking to expand their work in the area of greenhouse gas (GHG) measurement and management. EPA, headquartered in Washington, USA, is an independent agency of the United States and is led by a representative directly appointed by the President and approved by Congress.

In the last two decades scholars also focussed on the analysis of reputational indices at the environmental level, highlighting a positive relationship between environmental performance and Return on Assets (ROA), Return on Equity (ROE), and Return on Investment (ROI) (Erfle & Fratantuono, 1992; Hart & Ahuja, 1996). In particular, these studies consider the responsible behaviour and the level of environmental efficiency as key factors in determining better market performance (Derwall et al., 2005; Karpoff et al., 2005) as well as better profitability (see also Porter & Kramer, 2006; King & Lenox, 2001, 2002).

Reinhardt (1999) argues that reducing pollution may increase production efficiency, increase demand from environmentally sensitive consumers, discourage stakeholder activism, and allow a firm to attract better workers.

According to Coglianese and Nash definition (2001), an environmental management system consists of a collection of internal policies, assessments, plans, and implementation actions affecting the entire organizational unit and its relationships with the natural environment.

Darnall et al. (2008) state that both institutional pressures and capabilities may encourage the adoption of Environmental practices and improve business performance. To argue their hypothesis, they used data collected from a survey conducted by the Organisation of Economic Co-Operation and Development (OECD) and University researchers from Canada, France, Germany, Hungary, Japan, Norway, and the US. Their findings point out that firms that are incentivized to adopt more comprehensive environmental practices due to their complementary resources and capabilities, such as export orientation, employee commitment, and environmental R&D, observe greater overall business performance.

Ceteris paribus, in countries where environmental regulation is not effective, it is cheaper to operate than countries where strict environmental regulations result in fines, liabilities, legal, and administrative actions against polluters (Stewart, 1993).

On the other hand, Dowell et al. (2000) affirmed whether a stringent global corporate environmental standard represents a competitive asset or liability for multinational enterprises. They examine whether adopting a single stringent corporate environmental standard enhances firm value compared to those multinational companies defaulting to poorly enforced host country standards. According to this study, better firms appear to adopt higher environmental standards and pollute less. Moreover, Dowell et al. affirm the existence of a significant positive relationship between the firm market value captured by Tobin's Q and the level of environmental standards.

Other studies assessed whether it pays to be green. Cohen et al. (1995) show the presence of a strong correlation between environmental performance and firm profitability. Likewise, Hart and Ahuja (1996) provide evidence that efforts to reduce emission and prevent pollution drop to the "bottom line", measured by return on sales (ROS) and Return on Assets (ROA) within one to two years of initiation and that those firms with the highest emission levels stand to gain the most.

Again, Russo and Fouts (1997), using a sample of 243 firms, argue that the development of good environmental practices is positively related to economic performance. Industry growth moderates this link, showing that the environmental performance returns are higher for high-growth industries.

More recent studies suggest another key issue related to environmental performance as well as the environmental innovation. Innovation approach seems to lead to more efficient use of energy and material in environmental performance showing a positive effect on Firm Performance.

Ghisetti and Rennings (2014) point out that the relationship between Firm Performance and Environmental performance depends on the level of environmental innovation and the development of technological practices and mechanisms. This issue was so decisive that a specific ESG category provided by the Refinitiv methodology as well as "Environmental Innovation Score" was created. This category indicates companies' ability to reduce environmental costs creating new market opportunities through new technologies and eco-designed processes or products.

Turning to the social dimension, the literature provides different theories on the relationship between Corporate Social Performance and Firms' Performance. Stephen Brammer, Chris Brooks, and Stephen Pavelin (2006) argue that expenditures for some corporate social activities destroy shareholder value (Navarro, 1988), leading to a negative link between Social Scores and Firms' Returns.

Huselid (1995) analyses the relationship between Systems of High-Performance Work Practices[2] and Firm performance. Evidences highlight

[2] "High Performance Work Practices" consist into the comprehensive employee recruitment and selection processes, performance management systems, incentive compensation, employee involvement and performance.

that investments in High-Performance Work Practices are associated with lower employee turnover and greater productivity and corporate financial performance.

Also Richard et al. (2007) study the relationship between financial performance measured by Tobin's q and racial diversity in human resources. They used a data set of pooled time series and cross-sectional observations for American firms that took part in Fortune magazine's diversity survey, which was used to select the "50 Best Companies for Minorities" for the time span 1997–2002 and find evidence for a positive relationship between diversity and long-term financial performance.

However, more recent studies remark the importance of workers as a key factor. Faleye and Trahan (2011) focus on four measures of operating performance and firm value. They find that announcements of practices in favour of workspace are associated not only with positive abnormal stock returns but also to outperform compared to peer firms in terms of long-run operating results. In specific, the results show that the workforce category is associated with a significant increase in ROA.

Hillman and Keim (2001) instead use Market Value Added (MVA)[3] to explain shareholder value creation. According to the authors, this measure best captures the relative success of firms in maximizing shareholders' value through efficient allocation and management of scarce resources. They test whether there is a link between shareholder value, stakeholder management, and social issue participation. Starting from the preposition that building better relations with primary stakeholders like customers, suppliers, employees, and communities may lead firms to develop intangible and valuable assets, which could be a source of competitive advantage and consequently increase shareholder's wealth. This study involves data from S&P 500 firms and provides evidence that stakeholder management leads to improved shareholder value, while social issue participation is negatively associated with shareholder value.

Other studies point out that this positive relationship seems to be strongly influenced by the flexibility of the labour market (Edmans et al.,

[3] MVA is computed as the difference between market value and capital where market value refers to the equity market valuation of the company and capital is the debt and equity invested in the company. The authors choose to examine the change in MVA between one year and the following with the aim to more accurately reflect value changes caused by events in the previous year rather than total capitalization across time.

2017). In this context, the institutional environment in which managers and investors operate plays a crucial role.

Evidences highlight the huge importance of several aspects such as the recruitment, retention, and motivational benefits underlying employee satisfaction suggesting a greater value in countries where firms face fewer hiring and firing constraints. In contrast, in countries characterized by less flexible labour markets, returns are lower, leading to a downward shift in the marginal benefit curve of employee welfare spending.

At the same time, other Corporate Social Responsibility (CSR) schemes linked to the issue of gender diversity, environmental protection, and, more generally, a whole series of ethical issues are part of the same context.

In order to provide a more comprehensive literature framework, a recent study by Friede et al. (2015) collects more than 2,000 studies produced both by academics and investors. This survey, which represents to our knowledge, the main literature review study available up to that time, examines first studies and existing articles by including at the beginning studies based on vote counts, and then methodologically more accurate and sophisticated meta-analysis works, for a total of 3,718 studies[4] on the empirical relationship between ESG factors and Firm's Financial Performance. Based on a wide literature of academic research on the relationship between Financial Performance and ESG scores, authors highlight a positive and consistent ESG impact over time leading to interesting evidences for portfolio and non-portfolio studies, different asset classes, geographies, and E, S, and G categories.

Analysing individual categories of ESG Scores, the identification of material aspects is more immediate at the Environmental and Social level, while it seems to be trickier if we take into account Governance issues.

The consideration of ESG aspects and their impact on Corporate Financial Performances cannot disregard Shareholders' interests. The development of ESG practices inevitably impacts on shareholder value, where the interests of shareholders meet with those of company's Stakeholders. Specifically, the difference in corporate governance issues at the geographical level leads to identify three main reasons why the strength of corporate governance changes across countries:

[4] The authors argue that these 3,718 studies adjusted for overlap tend to reduce to a net number of more than 2,200 unique studies.

a. Ownership structure;
b. Shareholder orientation;
c. Assessing the level of governance in a company when it is isolated from the overall underlying institutional context.

According to Khan M. (2019), a good level of Corporate Governance determines not only a more efficient capital allocation but also greater preservation and growth of capital, necessary conditions to create sustainable value in the long term. On the other hand, non-sustainable companies are not able to provide long-term returns on savings such as a rate of return expected by shareholders, long-term employment, or sustainable tax revenues for social development. In line with this view, it seems that a low level of Corporate Governance is detrimental to both Shareholders and Stakeholders.

Other studies considered different variables of Corporate Governance. Mehran (1995) analysed the executive compensation structure of 153 companies operating in the manufacturing industry between 1979 and 1980. Estimates remark a positive effect on Firm Performance, as proxied by Tobin's Q and by ROA. In particular, the results show that Firm Performance is positively related both to the percentage of executive compensation equity-based and to the percentage of equity held by managers.

Furthermore, Mehran (1995) suggests that equity-based compensation is widely used in firms with more outside directors, and firms characterized by a larger percentage of shares owned by insiders or outside block-holders use less equity-based compensation.

Yermack (1996) instead measured the link between board size and firm value considering a panel of firms of annual *Forbes* magazine ranking of the 500 largest US public corporations based on sales, total assets, market capitalization, and net income. Using a sample of 452 large companies across eight years, between 1984 and 1991, the author finds an inverse relationship between board and firm value. More in detail, the association is characterized by a convex shape, suggesting that the largest fraction of lost value occurs as boards grow from small to medium size.

Core et al. (1999) aim to examine whether there is an association between the level of CEO compensation and the quality of firm's corporate governance and whether firms with better governance structures have better performance in the future, using a sample of 205 public-listed US companies operating in different industries over a three-year period. Firstly, they control for the standard economic determinants of compensations, firm's demand for a higher-quality CEO, and prior firm performance and risk. Then, the authors argue that board and ownership structure is associated with the level of CEO compensation. Considering the Board of Directors structure, Core et al. (1999) find that CEO compensation decreases if the board is composed of inside directors and is positively related to board size. Moreover, with regard to ownership structure, they point out that CEO compensation is a decreasing function of the CEO's ownership stake. Overall, the study suggests that firms with weaker governance structures are characterized by a high level of agency problems. Indeed, the firms with greater agency weaknesses provide greater CEO's compensation and consequently, these firms perform poorly.

More recent studies support the existence of a positive impact of governance policies considering the ownership and board structures and leverage (Beiner et al., 2006). Authors observe this relationship by constructing a broad corporate governance index. Using a sample of Swiss firms, they test the hypothesis of a positive relationship between corporate governance and firm valuation.

Other scholars as Ammann et al. (2011) used similar approaches to evaluate this issue. They construct two different corporate governance indices equally weighted related to the governance attributes and one index derived from a principal component analysis. They find a strong and positive relation between corporate governance and firm valuation in all three indices.

In last decades, an important issue that assumes growing relevance is disclosure. Within the academic literature, it has been observed that more and more companies (especially large firms) have intensified their efforts in reporting on ESG issues in order to justify their actions and improve firm's reputation. Khan et al. (2016) observe more than 7,000 companies reporting on their activities with respect to ESG issues, compared to a number of only 300 companies in 1996. In any case, despite the adoption of the Global Reporting Initiative (GRI) guidelines, the extent and quality of ESG-related disclosure appear heterogeneous (Ioannou & Serafeim, 2016).

Recent studies analysed the relationship between the role of disclosure related to the ESG performance and firm value (Fatemi et al., 2017). Estimates show that a good disclosure on ESG strengths tends to increase firm value, while weaknesses tend to decrease it. Despite this, if the ESG disclosure category is isolated, it is possible to observe a decrease in firm value. In particular, the study shows that in the presence of strengths related to ESG, a high level of disclosure allows for a less effective evaluation of the positive effect of these aspects. A possible explanation could be that the market might interpret excessive disclosure as an attempt by the firm to justify excessive investments in ESG activities.

On the other hand, it seems that disclosure also tends to weaken negative effects regarding the valuation of ESG issues. The reason could be found in the evidences because disclosure allows firms to justify their behaviour in explaining to investors the adequacy of their ESG operations and policies, or because firms convince investors that they have made credible commitments to modify their operations and thus overcome ESG weaknesses.

Conclusion

The awareness that natural resources are limited and the impact of human actions over time prompted institutions and firms to move towards a new paradigm more sustainably.

During the last two decades, academic literature has pointed out the need to adopt an approach more responsible aimed not only to maximize the shareholders' value but capable to consider contemporary other dimensions including environmental, social, and governance. However, researchers highlight the existence of several benefits for companies to adopt sustainable approach both in terms of operating performances and greater stock returns. The search for a relation between Environmental, Social, and Governance scores and Firm Financial Performance can be traced back to the beginning of 1970s. The increasing attention of investors determined particular interest by scholars and practitioners.

The comprehensive academic literature provided on this field will support the reader to capture the relevant findings of the following chapters. Starting to the overall academic literature on ESG, we seek to analyse the importance to develop Environmental, Social, and Governance practices and policies across (i) sectors including Oil and Gas, Automotive, and Pharmaceutical, (ii) countries, introducing also the perspective of Emerging Markets, and (iii) through the adoption of particular financial instruments such as green bonds.

References

Aktas, N., de Bodt, E., & Cousin, J. G. (2011). Do financial markets care about SRI? Evidence from mergers and acquisitions. *Journal of Banking and Finance, 35*(7): 1753–1761. Elsevier.

Ammann, M., Oesch, D., & Schmid, M. M. (2011). Corporate governance and firm value: International evidence. *Journal of Empiric Finance, 18*, 36–55.

Barth, M. E., Konchitchki, Y., & Landsman, W. R. (2013). *Cost of capital and earnings transparency*.

Bauer, R., & Hann, D. (2010). *Corporate environmental management and credit risk* (ECCE Working Paper). University Maastricht. The European Centre for Corporate Engagement.

Bebchuk, L. A., Cohen, A., & Ferrell, A. (2009). What matters in corporate governance? *Review of Financial Studies, 22*(2), 783–827.

Beiner, S., Drobetz, W., Schmid, M. M., & Zimmermann, H. (2006). An integrated framework of corporate governance and firm valuation. *European Financial Management, 2*, 249–283.

Bhojraj, S., & Sengupta, P. (2003). Effect of corporate governance on bond rating and yields: The role of institutional investors and outside directors. *Journal of Business, 76*(3), 455–475.

Bragdon, J. H., Jr., & Marlin, J. A. T. (1972). Is pollution profitable? Environmental virtue and reward: Must stiffer pollution controls hurts profits? *Risk Management Magazine*, pp. 1–8.

Brammer, S., Brooks, C., & Pavelin, S. (2006). Corporate social performance and stock returns: UK evidence from disaggregate measures. *Financial Management, 35*(3), 97–116.

Campbell, J. L. (2007). Why would corporations behave in socially responsible ways? An institutional theory of corporate social responsibility. *The Academy of Management Review, 32*, 946–967.

Chen, K. H., & Metcalf, R. W. (1980). The Relationship between Pollution Control Record and Financial Indicators Revisited. *The Accounting Review, 55*, 168–177.

Coglianese, C., & Nash, J. (Eds.). (2001). *Regulating from the inside: Can environmental management systems achieve policy goals?* Resources for the Future.

Cohen, M. A., Fenn, S., & Naimon, J. S. (1995). *Environmental and financial performance: Are they related?* Washington, DC: Investor Responsibility Research Center, Environmental Information Service.

Core, J. E., Holthausen, R. W., & Larcker, D. F. (1999). Corporate governance, Chief Executive Officer compensation, and firm performance. *Journal of Financial Economics, 51*, 371–406.

Cremers, K. J. M., & Ferrell, A. (2014). Thirty years of shareholder rights and firm value. *The Journal of Finance, 69*(3), 1167–1196.

Darnall, N., Henriques, I., & Sadorsky, P. (2008). Do environmental management systems improve business performance in an international setting? *Journal of International Management, 14*, 364–376.

Davis, K. (1973). The case for and against business assumption of social responsibilities. *Academy of Management Journal, 16*(2), 312–322.

Derwall, J., Bauer, R., Guenster, N., & Koedijk, K. (2005). The eco-efficiency premium puzzle. *Financial Analysts Journal, 61*(2): 51–63.

Dhaliwal, D. S., Li, O. Z., Tsang, A., & Yang, Y. G. (2011). Voluntary nonfinancial disclosure and the cost of equity capital: The initiation of corporate social responsibility Reporting. *American Accounting Association. The Accounting Review, 86*, 59–100.

Dowell, G., Hart, S. L., & Yeung, B. (2000). Do corporate global environmental standards create or destroy market value? *Management Science, 46*(8), 1059–1074.

Edmans, A., Li, L., & Zhang, C. (2017). *Employee satisfaction, labor market flexibility, and stock returns around the world* (No. w20300). National Bureau of Economic Research.

El Ghoul, S., Guedhami, O., Kwok, C. C. Y., & Mishra, D. R. (2011). Does corporate social responsibility affect the cost of capital? *Journal of Banking & Finance, 35*(9), 2388–2406.

El Ghoul, S., Guedhami, O., Kim, H., & Park, K. (2018). Corporate environmental responsibility and the cost of capital: International evidence. *Journal of Business Ethics, 149*(2), 335–361.

Elkington, J. (1994). Towards the sustainable corporation: Win-win-win business strategy for sustainable development. *California Management Review, 36*(2).

Elkington, J. (1997). *Cannibals with forks: The triple bottom line of 21st century business*. Capstone.

Erfle, S., & Fratantuono, M. J. (1992). *Interrelations among corporate social performance, social disclosure, and financial performance: An empirical investigation*. Dickinson College.

Faleye, O., & Trahan, E. A. (2011). Labor-friendly corporate practices: Is what is good for employees good for shareholders? *Journal of Business Ethics, 101*(1), 1–27.

Fatemi, A., Galum, M., & Kaiser, S. (2017). ESG performance and firm value: The moderating role of disclosure. *Global Finance Journal*.

Fisher-Vanden, K., & Thorburn, K. (2011). Voluntary corporate environmental initiatives and shareholder wealth. *Journal of Environmental Economics and Management, 62*(3): 430-445.

Freeman, R. E. (1984). *Strategic management. A stakeholder approach*. Pitman.

Friede, G., Busch, T., & Bassen, A. (2015). ESG and financial performance: Aggregated evidence from more than 2000 empirical studies. *Journal of Sustainable Finance & Investment*, pp. 210–233.

Friedman, M. (1970, September 13). The social responsibility of business is to increase its profits. *New York Times Magazine*.

Fulton, M., Kahn, B., & Sharples, C. (2012). *Sustainable investing: Establishing long-term value and performance*. Available at SSRN 2222740.

Ghisetti, C., & Rennings, K. (2014). Environmental innovations and profitability: How does it pay to be green? An empirical analysis on the German innovation Survey. *Journal of Cleaner Production, 75*(1), 106–117.

Gompers, P. A., Ishii, J., & Metrick, A. (2003). Corporate governance and equity prices. *The Quarterly Journal of Economics, 118*(1), 107–156.

Hart, S. L., & Ahuja, G. (1996). Does it pay to be green? An empirical examination of the relationship between pollution prevention and firm performance. *Business Strategy End Environment, 5*, 30–37.

Hillman, A. J., & Keim, G. D. (2001). Shareholder value, stakeholder management, and social issues: What's the bottom line? *Strategic Management Journal, 22*, 125–139.

Horvathova, E. (2010). Does environmental performance affect financial performance? A meta-analysis. *Ecological Economics, 70*, 52–59.

Huselid, M. A. (1995). The impact of human resource management practices on turnover, productivity, and corporate financial performance. *Academy of Management Journal, 38*(3), 635–672.

Ioannou, I., & Serafeim, G. (2016) *The consequences of mandatory corporate sustainability reporting: Evidence from four countries* (Harvard Business School Research Working Paper No. 11-100). Available on SSRN: https://ssrn.com/abstract=1799589

Karpoff, J. M., Lott, J. R., & Wehrly, E. W. (2005). The reputational penalties for environmental violations: Empirical evidence. *Journal of Law and Economics, 48*(2), 653–675.

Khan, M. (2019). Corporate governance, ESG, and stock returns around the world. *Financial Analyst Journal, 75*(4), 103–123.

Khan, M., Serafeim, G., & Yoon, A. (2016). Corporate sustainability: First evidence on materiality. *The Accounting Review, 91*(6), 1697–1724.

King, A., & Lenox, M. (2001). Does it really pay to be green? An empirical study of firm environmental and financial performance: An empirical study of firm environmental and financial performance. *Journal of Industrial Ecology, 5*(1), 105–116.

King, A., & Lenox, M. (2002). Exploring the locus of profitable pollution reduction. *Management Science, 48*(2), 289–299.

Konar, S., & Cohen, M. A. (2001). Does the market value environmental perfomance. *Review of Economics and Statistics, 83*, 281–289.

Lima, B. F., & Sanvicente, A. Z. (2013). Quality of corporate governance and cost of equity in Brazil. *Journal of Applied Corporate Finance, 25*(1), 72–80.

Margolis, J. D., Elfenbein, H. A., & Walsh, J. P. (2007). *Does it pay to be good? A meta-analysis and redirection of research on the relationship between corporate social and financial performance*. Ann Arbor 1001 (48109-1234).

Mehran, H. (1995). Executive compensation structure, ownership, and firm performance. *Journal of Financial Economics, 38*(2), 163–184.

Navarro, P. (1988). Why do corporation give to charity? *The Journal of Business, 61*(1), 65–93.

Orlitzky, M., Schmidt, F. L., & Rynes, S. L. (2003). Corporate social and financial performance: A meta-analysis. *Organization Studies, 24*, 403–441.

Porter, M. E., & Kramer, M. R. (2006). Strategy & society: The link between competitive advantage and corporate social responsibility. *Harvard Business Review, 84*, 78–92.

Reinhardt, F. (1999). Market failure and environmental policies of firms. *Journal of industrial Ecology, 3*(1).

Reverte, C. (2012). The impact of better corporate social responsibility disclosure on the cost of equity capital. *Corporate Social Responsibility and Environmental Management, 19*(5), 253–272.

Richard, O. C., Murthi, B. P. S., & Ismail, K. (2007). The impact of racial diversity on intermediate and long-term performance. The moderating role of environmental context. *Strategic Management Journal, 28*, 1213–1233.

Russo, M. V., & Fouts, P. A. (1997). A resource based perspective on corporate environmental performance and profitability. *Academy of Management Journal, 40*(3), 534–559.

Schauten, M., & Van Dijk, D. J. C. (2011). Corporate Governance and the Cost of Debt of Large European Firms. ERIM Report Series Research in Management. No. ERS-2010-025-F&A.

Soppe, A., Schauten, M., Soppe, J., & Kaymak, U. (2011). Corporate social responsibility reputation (CSRR): Do companies comply with their raised CSR expectations? *Corporate Reputation Review, 14*(4), 300–323.

Stewart, R. (1993). Environmental regulation and international competitiveness. *Yale Law Journal, 102*, 2039–2106.

Van Beurden, P., & Gössling, T. (2008). The worth of values—A literature review on the relation between corporate social and financial performance. *Journal of Business Ethics, 82*(2), 407–424.

Waddock, S. A., & Graves, S. B. (1997). The corporate social performance—Financial performance link. *Strategic Management Journal, 18*(4), 303–319.

Yermack, D. (1996). Higher market valuation of companies with a small board of directors. *Journal of Financial Economics, 40*(2), 185–211.

PART II

ESG, Cost of Equity and Firm' Performance: Empirical Evidences Across Industries Around the World

CHAPTER 7

ESG, COE and Profitability in the Oil and Gas Sector

Carlo Bellavite Pellegrini, Raul Caruso, and Marco Seracini

Introduction

It is nowadays clear that ESG issues concern most of the business choices in a transversal way to all sectors of activity, and there is no doubt that the Oil and Gas sector is the most impacted one by the need for an unavoidable energy transition towards the production of clean energy, which has to be in line with the objectives of containing greenhouse gas emissions (GHG).

C. Bellavite Pellegrini · R. Caruso
Department of Economic Policy, Catholic University of the Sacred Heart, Milan, Italy
e-mail: carlo.bellavite@unicatt.it

C. Bellavite Pellegrini · R. Caruso (✉) · M. Seracini
Research Centre of Applied Economics, Catholic University of the Sacred Heart, Milan, Italy
e-mail: raul.caruso@unicatt.it

© The Author(s), under exclusive license to Springer Nature Switzerland AG 2022

C. B. Pellegrini et al. (eds.), *Climate Change Adaptation, Governance and New Issues of Value*, Palgrave Studies in Impact Finance,
https://doi.org/10.1007/978-3-030-90115-8_7

The Oil and Gas industry has, as it is well known, peculiar aspects, such as: (a) It operates on natural resources with high environmental impacts and risks; (b) It has an international dimension; (c) It is strongly influenced by geopolitical factors; (d) It has a high technological content and it requires high investments; (e) It has a long-term production cycle (even over 30 years) and (f) It is influenced by the performance of financial markets and commodities.

Therefore, for oil companies, especially the larger ones, the transformation from Big Oil to Big Energy represents a complex, sensitive and long-term process, with choices on sustainable investments to be planned considering their own economic and financial balance. Investors have to monitor this process very carefully, balancing short-term performance expectations with the goal of ensuring a growth in value over time to their assets (Filippetti, 2019; Tamburi Investment Partners, 2017). Which is to say, in the Oil and Gas sector—and especially in the Upstream sector—ESG-oriented policies and investments.

For these reasons, this chapter empirically explores Oil and Gas sector, trying to investigate the effect of ESG Scores on (1) Cost of equity (COE) and (2) Firm's profitability (FP) for a sample of 182 operating global companies belonging to this industry between 2002 and 2018.

As we already know, the ESG Scores are synthetic indicators which are based respectively on environmental, social and governance aspects and practices which influence and shape the behaviour of firms.

Although corporate finance has historically researched about the determinants of stock returns and modelling future yields, recently corporate governance has focussed its attention on measuring the impact of non-financial information on listed companies' financial performance. This field of study has become more relevant over time due to the increasing attention of investors.

In fact, a large growing literature is nowadays investigating to what extent sustainable strategies affect both firm's performance and value.[1] Needless to say, the challenge is to verify whether considering sustainability, environmental and social issues also payoffs in terms of performance and added value to the firm.

Very briefly, there is a recently blossoming literature on both theoretical and empirical evidences related to Sustainability Performance and

[1] For an in-depth analysis on both theoretical and empirical literature, see Chapter 6.

the Cost of Equity (El Ghoul et al., 2018; Sharfman & Fernando, 2008; Suto & Takehara, 2017). Whether it is reasonable to say that ESG strategies of firms do contribute to the establishment of a more sustainable business context as envisioned in Waddock (2017), there are substantial doubts about the role of ESG in shaping both profitability and firm value (Dowell et al., 2000; Hart & Ahuja, 1996; Konar & Cohen, 2001; Lee et al., 2018). Some of the recent studies supporting the argument that a better Sustainability Performance generates a reduction in the Cost of Equity (Dhaliwal et al., 2014; Gupta, 2018; Matthiesen & Salzmann, 2017).

The sample of analysis is composed of firms of different dimensions. The majority of the firms composing the sample are characterized by size dimensions that range from 1 to 60 billion US dollars of market capitalization, meanwhile a smaller part, have a market capitalization that exceeds 60 billion US dollars. Analysing the sample from a geographical point of view, the greatest part of them is headquartered in the United States, Canada and Continental Europe.

While other scholars use well-known models such as the CAPM or the Fama and French Model, the added value of our work lays in the use of implied cost of equity measured according to Easton Model (Easton, 2004).

More specifically, in the first analysis we estimate firms' ex-ante cost of equity adopting Easton Model (2004), which expresses the share price in terms of one-year-ahead expected dividend per share and one and two-year-ahead expected earnings per share. The ESG Scores used for this study are drawn from Thomson Reuters Datastream,[2] which considers more than 180 industry-relevant sustainability variables that successively are aggregated into ten main E, S, and G components. By employing a fixed effect regression model and a parsimonious set of control variables, we show that firms with higher ESG Scores exhibit cheaper equity financing. In particular, our findings suggest that for a ten percent increase in the ESG Overall Score, the cost of equity of firms declines by 134 bps. Among other findings we underline that this relationship is not linear, instead, it has a U-shaped form. This means that greater attention towards ESG topics is beneficial for firms until they

[2] Datastream considers more than 180 industry-relevant sustainability variables that successively are aggregated into ten main E, S and G components.

reach a "threshold" in terms of size measured by total assets. Afterwards the relationship becomes positive.

For the second analysis instead, we consider Return on Assets as a proxy for firm's profitability and use the same dataset as in the previous analysis. We show that better ESG performance is negatively related with Return on Assets. In specific, in the presence of a ten percent increase in the overall ESG Score the Return on Assets of firms in our dataset declines by 0.45%. The same non-linear, U-shaped form, relationship persists also in the profitability analysis.

The obtained results of this empirical research are in line with the literature, supporting the argument that a better Sustainability Performance generates a reduction in the Cost of Equity (Dhaliwal et al., 2014; Gupta, 2018; Matthiesen & Salzmann, 2017). For both analyses, COE and FP, we employ a semi-logarithmic fixed effect regression model implementing various robustness tests in order to check whether the same effects hold in more recent times (2010–2018/2019) when the availability of data is greater and considering different firms' size.

Data and Empirical Strategy

Given the peculiarities of the Oil and Gas sector and its extended exposure towards ESG topics, we are interested to check whether the scores attributed to the ESG profile of firms is reflected on their cost of equity and in their profitability. To do so we construct a dataset composed of Oil and Gas producing firms that operate worldwide and compute the following two analyses:

A. ESG scores and Cost of Equity
B. ESG scores and Firms Profitability

The time period considered spans from 2002 to 2018 and is chosen in order to incorporate the largest and most reliable set of ESG scores. Several criteria were applied in creating the dataset: (i) only firms whose ESG scores were available for more than five years were considered and (ii) these firms ought to have analyst coverage in order to obtain up to two years of forecasted earnings and dividend per share for the valuation models. This selection process led to the construction of a panel

Table 7.1 Sample distribution by market capitalization (billion US$)

Market capitalization	*Number of firms*
0–1	42
1–5	53
5–10	22
10–20	20
20–30	12
30–60	18
60–100	8
100–200	3
+200	4
Total number of firms	**182**

This table shows the sample distribution based on firms' size measured by market capitalization. The largest numbers of firms in the sample belong to the range of market capitalization of 1–5 billion US$

Source Thomson Reuters Datastream; Authors' elaboration

composed of 182 firms of different dimensions operating in the Oil and Gas sector which was used for both analyses (Tables 7.1 and 7.2).

The ESG scores are taken from the data provider Thomson Reuters Datastream which captures and calculates over 400 company-level

Table 7.2 Sample distribution by geographic area

Geographic area	*Number of firms*
North America and Canada	86
Continental Europe	23
Asia	18
Pacific Asia	18
United Kingdom	12
Oceania	10
Scandinavian Europe	9
Latin America	5
Middle East	1
Total number of firms	**182**

This table presents the composition of the sample based on geographic area. The majority of firms in the sample are represented by North American and Canadian firms, followed by Continental European companies and Asian ones

Source Thomson Reuters Datastream; Authors' elaboration

measures, of which they select a subset of 178 most comparable and relevant fields to power the overall company assessment and scoring process. The underlying measures are based on considerations around comparability, data availability and industry relevance. They are grouped into ten categories, weighted proportionally to the count of measures within each category formulates the final ESG score, which reflects the company's ESG performance, commitment and effectiveness based on publicly reported information. These scores range from 0 to 100 where a greater score means greater commitment towards ESG topic. The categories that compose the Environmental Score are: (1) Resource Use score, measures the commitment of a company to reduce the use of energy, water and materials and to introduce more eco-efficient solutions by enriching the supply chain management; (2) Emissions Reduction score, represents the commitment and the effectiveness of the reduction of environmental emission in the operational processes and (3) Innovation score, takes account of the company's capacity to reduce the environmental costs for its customers by creating new market opportunities through the use of new environmental technologies and eco-friendly designed products. The categories composing the Social Score are: (1) Workforce score, represents a company's commitment to guarantee job satisfaction, a healthy and safe workplace, supporting diversity and equal opportunities; (2) Human Rights score, quantifies a company's effectiveness in respecting the fundamental human rights conventions; (3) Community score, represents the attempts of the firm in being a good citizen, contributing into public health and respecting business ethics and (4) Product responsibility score, reflects company's capacity to guarantee quality goods and services integrating the customer's health and safety, integrity and data privacy. The Governance Score captures: (1) Management score, represents a company's commitment to follow best practice corporate governance principles; (2) Shareholders score, reflects the effectiveness regarding equal treatment of shareholders and the use of anti-takeover devices and (3) CSR Strategy score, comprehends the practices a company applies with the scope of integrating the economic-financial, social and environmental dimensions into its decision-making process (Table 7.3).

In addition, variables to control for financial peculiarities, which differ moving from the COE analysis and the FP analysis were used. Regarding the COE analysis which considers the cost of equity calculated using

Table 7.3 Descriptive statistics ESG scores of firms composing the sample

ESG score components	*Obs*	*Mean*	*Std. Dev*	*Min*	*Max*
ESG overall score	2,114	56.395	17.426	1.098	92.915
Resource use score	2,114	58.665	27.509	0.224	99.797
Emissions score	2,114	58.657	27.571	0.276	99.798
Environmental Innovation Score	2,114	52.401	24.660	0.202	99.795
Human rights score	2,114	55.998	25.682	14.722	99.747
Community score	2,114	55.708	29.304	14.722	99.798
Workforce score	2,114	58.346	27.583	0.202	99.796
Product responsibility score	2,114	55.337	26.650	0.234	99.798
Management Score	2,114	55.497	29.110	0.505	99.950
Shareholders score	2,114	54.220	27.526	0.526	99.924
CSR strategy score	2,114	58.830	27.470	0.051	99.843

Source Thomson Reuters Datastream; Authors' elaboration

Table 7.4 Descriptive statistics independent and control variables for both analysis

	Obs	*Mean*	*Min*	*Max*
COE	3.074	0.113	0	1.717
ROA	2.921	0.046	−2.428	0.837
Total assets[a]	2.937	28.401	0.199	440.901
Leverage	2.937	0.702	−80.904	122.335
Market to book value	2.719	2.134	−178.940	81.780
Long-term growth estimate	2.016	14	−183	177

[a]Millions of US dollars
Source Thomson Reuters Datastream; Authors' elaboration

Easton's model as an independent variable,[3] the control variables are as follows: (i) Firms size measured by total assets; (ii) Long-term growth rate; (iii) Market to book value and (iv) Time trend variable. For what concerns the FP analysis which uses Return on Assets (ROA) as a proxy for profitability instead, the control variables employed are: (i) Firms size measured by total assets; (ii) Financial leverage; (iii) Market to book value and (iv) One-year lagged profitability measure (ROA) (Table 7.4).

[3] See Appendix A of this chapter.

ESG Scores and Cost of Equity: COE Analysis

In order to analyse the relationship between ESG scores and firm's Cost of Equity, we employ the following fixed effect regression model:

$$\begin{aligned}\text{Cost of Equity}_{it} &= \alpha + \beta_1 \log \text{ESG}_{it} + \beta_{n+1} \log \text{Control Variables}_{it} \\ &+ \text{Time Trend}_{it} + \varepsilon_{it}\end{aligned}$$

The choice to consider the implied cost of equity is supported by El Ghoul et al. (2011) and Dhaliwal et al. (2011, 2014) who show that both the standard single-factor model and the Fama and French (1993) three-factor model provide poor proxies for the cost of equity. Hail and Leuz (2011) and Chen et al. (2009) argue that the implied cost of capital approach is particularly useful because it makes an explicit attempt to isolate cost of capital effects from growth and cash flow effects, as occurs for the more generally used ex-post models based on realized returns. The output of Easton model (2004) represents the final measure of COE in our analysis. This model allows the share price to be expressed in terms of one-year-ahead earnings per share forecasts. The explicit forecast horizon is set to two years, after which forecasted abnormal earnings are assumed to grow in perpetuity at a constant rate. The model requires positive one-year-ahead and two-year-ahead earning forecast.

The valuation equation of the Easton Model (2004) is given by:

$$P_0 = \frac{\text{eps}_2 + \text{COE} * \text{dps}_2 - \text{eps}_1}{\text{COE}^2}$$

$$\text{COE} = \frac{\sqrt{\text{eps}_{2+}\text{COE} * \text{dps}_2 - \text{eps}_1}}{P_0}$$

where eps_1 and eps_2 are the forecasted values of earnings per share in time $t + 1$ and $t + 2$ and dps_2 is the forecasted dividend per share in time $t + 2$. The data employed in the above equations are forecast data obtained by I.B.E.S. database, part of Thomson Reuters Datastream.

The above equation generates two results, only the positive outputs were considered and subsequently implemented into to regression model. All the necessary diagnostic tests were taken, confirming the fixed effect regression model as the best fit. The obtained results are as presented in Table 7.5.

Table 7.5 Results of COE analysis

	ESG	*Environmental*			*Social*				*Governance*		
	ESG score	*Resource use score*	*Emissions score*	*Environmental innovation score*	*Workforce score*	*Human rights score*	*Community score*	*Product responsibility score*	*Management score*	*Shareholders score*	*CSRstrategy score*
Model	*5.1*	*5.2*	*5.3*	*5.4*	*5.5*	*5.6*	*5.7*	*5.8*	*5.9*	*5.10*	*5.11*
Beta coefficient	**−0.14112***	−0.0171	−0.0091	−0.0772	**−0.0426****	0.0242	0.0183	**−0.0533***	0.0046	−0.0429	- **0.0626****
	(0.0778)	(0.0204)	(0.0149)	(0.0953)	(0.0190)	(0.0792)	(0.0369)	(0.0320)	(0.0275)	(0.0426)	(0.0254)
Total assets	−0.6198	−0.6042	−0.6096	−0.7632	−0.5564	−0.6143	−0.6160	−0.7010	−0.6189	−0.5726	−0.5651
	(0.4910)	(0.4972)	(0.4875)	(0.5683)	(0.4902)	(0.4889)	(0.4979)	(0.5009)	(0.4972)	(0.4671)	(0.4872)
Total $assets^2$	0.0196	0.0189	0.0191	0.0240	0.0174	0.0191	0.0193	0.0220	0.0194	0.0178	0.0178
	(0.0158)	(0.0160)	(0.0157)	(0.0182)	(0.0157)	(0.0157)	(0.0159)	(0.0160)	(0.0159)	(0.0150)	(0.0156)
Long-term growth	0.0569	0.0579	0.0577	0.0556	0.0542	0.0571	0.0569	0.0571	0.0580	0.0591	0.0613
	(0.0351)	(0.0350)	(0.0351)	(0.0357)	(0.0353)	(0.0357)	(0.0350)	(0.0350)	(0.0354)	(0.0347)	(0.0356)
Market to book value	0.0412	0.0381	0.0388	0.0330	0.0423	0.0389	0.0381	0.0417	0.0381	0.0418	0.0405
	(0.0400)	(0.0396)	(0.0397)	(0.0406)	(0.0397)	(0.0395)	(0.0394)	(0.0406)	(0.0398)	(0.0402)	(0.0401)
No. observations	699	699	699	699	699	699	699	699	699	699	699

***, **, * stand for statistical significance at 1%, 5% and 10%, respectively

This table presents the results obtained by employing a half-logarithmic fixed effect regression model with clustered robust errors. Each column contains the output of a regression that considers Cost of Equity (COE) as dependent variable and the component of ESG score as main independent variable. The clustered robust errors are presented in brackets. The control variables employed into the model are: Total assets; Long-Term Growth, Market to Book Value; Time Trend. The employed models are in a half-logarithmic form: all the control variables and the main explanatory variable (ESG score) are in logarithmic form meanwhile the dependent variable (COE) is not.

Source Thomson Reuters Datastream; Authors' elaboration

As the results show, we detect a statistically significant negative association between COE and ESG. Overall Score there is a 134.5 bps[4] reduction in the cost of equity of firms that operate in the Oil and Gas sector. Looking more in detail into the components of each factor we obtain robust results for the Social and Governance factors. In specific there is a decrease of around 60 bps and 40.6 bps for every ten percent increase in the Product Responsibility score and in the Workforce score, respectively. Regarding the Governance factor we find a 50.8-bps decrease in COE deriving from greater scores of CSR Strategy.

Among the control variables we included the quadratic term of the firm's size measure (Total Assets) which plays the role of a simple robustness test seeking for a non-linear relationship between the ESG scores and the dependent variables. Results are puzzled and suggest the existence of non-linearities in the relationship we are investigating. In specific, the relation between COE and ESG scores is characterized by a U-shaped form. The impact of greater ESG performance is negatively related to the cost of equity of firms until the size of the firm reaches a certain level, afterwards the relation becomes positive.

To confirm the robustness of our results we computed various robustness test by (i) decreasing the years of observation, (ii) removing the 20 largest companies and (iii) removing the 20 smallest companies in the sample (Tables 7.7, 7.8, 7.9 in Appendix B). The results obtained are in line with our findings confirming once again the negative association between the variables.

ESG Scores and Firms Profitability: FP Analysis

In order to analyse the relation between the profitability and ESG performance we use a dataset that contains the same firms as previously and employ the following half-logarithmic fixed effect regression model:

$$
\begin{aligned}
\mathrm{ROA}_{it} = {} & \alpha + \beta_1 \log \mathrm{ESG\,Score}_{it} + \beta_{n+1} \log \mathrm{Control\,Variables}_{it} \\
& + \beta_{n+2}\mathrm{ROA}_{i,t-1} + \varepsilon_{it}
\end{aligned}
$$

[4] Since the regressions are in a half-logarithmic form, the results are read as follows: Quantitative effect = β*log (1.10).

The dependent variable is the Return on Asset (ROA), computed as the ratio between net income and total assets. We also employ a parsimonious set of control variables established in the existing literature: size variable, leverage variable, market performance measure and past profitability.

Table 7.6 reports the results. The overall ESG Score does exhibit a statistically significant negative association with ROA. This means that for a ten percent variation of the ESG score, the profitability of the firm measured by ROA reduces by 0.45%. Analysing in detail each component, from the Environmental factor we find evidence that the Resource Use score is negatively related to firm's profitability. The same type of relation is found also for the subcategories composing the Social factor: Community score and Workforce score. The negative association persists also for the Governance factor captured by CSR Strategy score and Shareholders score.

Repeating the same approach as in the COE analysis, we add the quadratic term of the size measure and we obtain the same non-linear relationship in a U-shaped form.

Also, in the FP analysis we compute the robust test by reducing the observation years into 2010–2018 and excluding from the sample firstly the 20 biggest firms and successively the 20 smallest firms. The robustness of our model is confirmed since we obtained the same type of relationship between the ESG scores and the profitability measure (see Tables 7.10, 7.11, 7.12 in Appendix C). It is interesting to note that when we consider the sample which excludes the 20 smallest firms, we observe a non-linear relationship but, in this case, it has an inverse U-shaped form suggesting that the efficiency of these sustainability measures is strictly related with firm's size.

Conclusions

This work focussed on the impact of ESG scores on Cost of Equity and Firms' Profitability of a panel of 182 global listed firms operating in the Oil and Gas sector over the period between 2002 and 2018. Our main findings highlight that:

i. The overall ESG score is negatively associated with Cost of Equity of firms, measured by the Easton Model. When the ESG score increases by ten percent the Cost of Equity decreases by 134 bps.

Table 7.6 Results of FP analysis

	ESG	*Environmental*			*Social*				*Governance*		
	ESG score	*Resource use score*	*Emissions score*	*Environmental innovation score*	*Workforce score*	*Human rights score*	*Community score*	*Product responsibility score*	*Management score*	*Shareholders score*	*CSRstrategy score*
	6.1	*6.2*	*6.3*	*6.4*	*6.5*	*6.6*	*6.7*	*6.8*	*6.9*	*6.10*	*6.11*
Coefficient	**−0.0479***** (0.0141)	**−0.0094**** (0.0038)	−0.0072 (0.0051)	−0.0003 (0.0032)	**−0.0071*** (0.0037)	−0.0094 (0.0104)	**−0.0094**** (0.0037)	−0.0036 (0.0055)	−0.0059 (0.0063	**−0.0061**** (0.0029)	**−0.0084**** (0.0042)
Total assets	−0.0657 (0.1021)	−0.0783 (0.1005)	−0.0742 (0.0978)	−0.0803 (0.1003)	−0.0726 (0.1009)	−0.0842 (0.1007)	−0.0745 (0.0997)	−0.0841 (0.1017)	−0.0771 (0.1030)	−0.0715 (0.0985)	−0.0752 (0.1012)
Total assets^2	0.0018 (0.0031)	0.0021 (0.0306)	0.0019 (0.0030)	0.0021 (0.0031)	0.0002 (0.0031)	0.0022 (0.0031)	0.0019 (0.0030)	0.0022 (0.0031)	0.0020 (0.0031)	0.0018 (0.0030)	0.0019 (0.0031)
Market to book value	0.0456 (0.0112)	0.0461 (0.0112)	0.0465 (0.0112)	0.0464 (0.0112)	0.0463 (0.0112)	0.0461 (0.0113)	0.0462 (0.0117)	0.0465 (0.0112)	0.0465 (0.0112)	0.0465 (0.0112)	0.0460 (0.01129
Leverage	−0.0386 (0.0076)	−0.0388 (0.0077)	−0.0391 (0.0078)	−0.0395 (0.0078)	−0.0391 (0.0078)	−0.0390 (0.0077)	−0.0388 (0.0078)	−0.0391 (0.0078)	−0.0392 (0.0078)	−0.0392 (0.0078)	−0.0391 (0.0078)
ROA (*t* − 1)	0.0665 (0.0554)	0.0731 (0.0568)	0.0739 (0.0577)	0.0774 (0.0569)	0.0745 (0.0562)	0.0767 (0.0569)	0.0727 (0.0571)	0.0767 (0.0570)	0.0766 (0.0566)	0.0764 (0.0567)	0.0748 (0.0564)

	ESG	*Environmental*			*Social*				*Governance*		
	ESG score	*Resource use score*	*Emissions score*	*Environmental innovation score*	*Workforce score*	*Human rights score*	*Community score*	*Product responsibility score*	*Management score*	*Shareholders score*	*CSRstrategy score*
	6.1	*6.2*	*6.3*	*6.4*	*6.5*	*6.6*	*6.7*	*6.8*	*6.9*	*6.10*	*6.11*
No. observations	1,926	1,926	1,926	1,926	1,926	1,926	1,926	1,926	1,926	1,926	1,926

***, **, * stand for statistical significance at 1%, 5% and 10%, respectively

This table presents the results obtained by employing a half-logarithmic fixed effect regression model with clustered robust errors. Each column contains the output of a regression that considers Return on Assets (ROA) as a dependent variable and the component of ESG score as a main independent variable. The clustered robust errors are presented in brackets. The control variables employed into the model are: Total assets; Financial Leverage, Market to Book Value; Past Profitability (ROA lagged one period). The employed models are in a half-logarithmic form: all the control variables and the main explanatory variable (ESG score) are in logarithmic form meanwhile the dependent variable (ROA) and the past profitability measure are not

Source Thomson Reuters Datastream; Authors' elaboration

ii. Same inverse association holds for Workforce score, Product Responsibility score and CSR Strategy score.
iii. We find that these negative associations are characterized by a non-linear U-shaped relationship.
iv. The firms' profitability measured by ROA is negatively related to better performance of ESG scores.
v. We obtain statistically significant results for the overall ESG Score suggesting that for a ten percent increase in the ESG score, there is a decrease in the ROA of around 0.45%.
vi. Statistically significant results derive from the subcategories of the Social factor (Community score and Workforce score) and Governance factor (CSR Strategy score and Shareholders score) which show a negative association with ESG scores.
vii. The relationship between firms' profitability and ESG scores is non-linear and is characterized by a U-shape form.

   Our findings support arguments in the literature that firms with better ESG performance have higher value and lower risk (Chen et al., 2009; El Ghoul et al., 2011, 2018; Hail & Leuz, 2011) and in the same time highlight some peculiarities deriving from industry-level factors (Gregory et al., 2016; Reverte, 2012). In term of future research, we would like to expand our analysis in other sectors in order to check whether the degree of materiality of ESG scores changes among different industries and different value chains. Moreover, notwithstanding the relatively short period taken into analysis and the choice of ROA as a proxy for firm's profitability, the use of other corporate variables like Tobin Q, may eventually add innovative evidence in the dynamic of Oil and Gas industry.

Generally speaking, only in the last part of the period considered in this study (from 2002 to 2018) the awareness of the urgent need for ESG-oriented choices in the Oil and Gas sector has emerged.

The turning point was undoubtedly 2015, with the Paris agreements on climate change and the signing, by 193 countries, of the UN Agenda for Sustainable Development: an action plan that has defined the three dimensions of development (economic, environmental and social) in 17 Sustainable Development Goals (SDGs) to be achieved by 2030.

All the majors in the sector are now committed to give an increasingly rapid impulse to ESG-oriented policies and investments, and the need to find a fair balance between the interests of shareholders and stakeholders represents the decisive challenge for the future of these companies.

It is particularly interesting in the case of ENI, one of the majors in the sector, which, since 2014, has embarked on a process of transformation of its business model through a decarbonization process oriented towards carbon neutrality in the long term, with huge investment plans in in the diffusion of renewable sources.

This new approach has found an innovative disclosure tool in the adoption of a long-term strategic plan, from 2020 to 2050, announced to the market in February 2020, which combines the goals of continuous development in a rapidly evolving market, such as the Energy one, with a significant reduction in the carbon footprint of the portfolio. It is a plan with stated objectives, which are punctually defined and articulated in an accurate timeline, and, therefore, measurable and verifiable.

ENI is an example of how the main companies in the Oil and Gas sector are finalizing their investments towards environmentally sustainable objectives, which, however, must be combined with the economically sustainable ones, which, at least in the short and medium term, represent the traditional business model.

Therefore, it arises the need to monitor and detect what the company performance resulting from this new scenario will be in the near future, which will certainly be subject to careful evaluation by investors. To allow the market the possibility of evaluating the correlations in a homogeneous way between ESG scores and financial and economic performance, it will be necessary to arrive at uniform metrics also in terms of ESG.

A recent step forward in this direction is represented by the signing, in September 2020, by 61 leaders of the most important companies in the world, including members of the World Economic Forum (WEF), of the fundamentals of the "Stakeholder Capitalism Metrics" issued by the International Business Council (IBC). These metrics offer a set of universal and comparable information focussed on people, planet and governance, about which companies, investors and all the various stakeholders can rely on, regardless of the sector or country in which they operate.

Further interesting development in the field of metrics is the one proposed by Mark Kramer in a recent publication,[5] in which, in reiterating the need for ESG indexes, in their calculation, not to be completely

[5] Kramer (2020) Hybrid metrics—Connecting shared value to shareholder value. *The Harvard Business Review*.

disconnected from the purely financial aspects of corporate performance, he indicates that the most suitable tool for this purpose is the use of "hybrid metrics" that can combine the social and environmental impact of companies with their standard financial performance measures.

Appendix A

Eatson Model

The model is based on the recognition of the central role of short-term forecasts of earnings in valuation. The roles of (1) forecasts of next period's accounting earnings, (2) forecasts of accounting earning two-period ahead and (3) expected accounting earnings beyond the two-year forecast horizon. The model shows how the difference between accounting earnings and economic earnings characterizes the role of accounting earnings in valuation.

Starting with the no-arbitrage assumption:

$$P_0 = (1 + R)^{-1}[P_1 + \mathrm{DPS}_1] \tag{7.1}$$

where:

P_0 = current, date $t = 0$, price per share;

P_1 = expected, date $t = 1$, price per share;

DPS_1 = expected dividends per share, at date $t = 1$;

R = expected rate of return and $R > 0$ is a fixed constant. Adding and subtracting capitalized accounting yields:

$$P_0 = \frac{\mathrm{EPS}_1}{R} - \frac{\mathrm{EPS}_1}{R} - (1 + R)^{-1}[P_1 + \mathrm{DPS}_1] \tag{7.2}$$

If expected accounting earnings EPS_1 is equal to economic earnings ($P_0 * R$), then the term in the brackets must equal to zero—in other words, next period's expected earnings are sufficient for valuation. However, if EPS_1 does not equal economic earnings then valuation based on accounting earnings requires forecasts beyond the next period.

$$P_1 = \frac{\mathrm{EPS}_2}{R} - \frac{\mathrm{EPS}_2}{R} - (1 + R)^{-1}[P_2 + \mathrm{DPS}_2] \tag{7.3}$$

Substituting Eq. (7.3) into Eq. (7.2) yields:

$$P_0 = \frac{EPS_1}{R} - (1+R)^{-1}\text{agr}_1 + (1+R)^{-2}R^{-1}[R * \text{DPS}_2 - (1+R)\text{EPS}_2] + (1+R)^{-2}P_2 \quad (7.4)$$

where

$$\text{agr}_1 = [\text{EPS}_2 + R * \text{DPS}_1 - (1+R)\text{EPS}_1] \quad (7.5)$$

is the expected abnormal growth in accounting earnings. This abnormal growth in earnings reflects the effects of generally accepted accounting practices that lead to a divergence of accounting earnings from economic earnings. If EPS_1 and EPS_2 were equal to economic earnings, then agr_1 would be zero and the ratio of expected earnings to price would be equal to the expected rate of return.

The valuation role of expected accounting earnings beyond the two-year forecast horizon may be seen by substituting for P_2, P_3, P_4, *etc.*, in Equation (7.5) to yield:

$$P_t = \frac{\text{EPS}_1}{R} + R^{-1}\sum\nolimits_{t=1}^{\infty}(1+R)^{-1}\text{agr}_t \quad (7.6)$$

Equation (7.6) shows that the present value of the agr_t sequence explains the difference between price and capitalized expected earnings. Equation (7.6) may be modified to accommodate a finite forecast horizon by defining a perpetual rate of change in abnormal growth in earnings (Δagr) beyond the forecast horizon. If earnings forecasts are available for two periods, Equation (7.6) may be written as:

$$P_0 = \frac{\text{EPS}_1}{R} + \frac{\text{agr}_1}{(R - (R - \Delta\text{agr}))} \quad (7.7)$$

where:

$$\Delta\text{agr} = \left(\frac{\text{agr}_{t+1}}{\text{agr}}\right) \quad (7.8)$$

Considering the special case $\Delta\text{agr} = 0$, meaning that $\text{agr}_1 = \text{ag}r_2 = \ldots$, from Eq. (7.7) we have:

$$P_0 = \frac{\text{EPS}_2 + R * \text{DPS}_2 - \text{EPS}_1}{R^2}$$

$$R = \sqrt{\frac{[\mathrm{EPS}_2 + R * \mathrm{DPS}_2 - \mathrm{EPS}_1]}{P_0}}$$

Appendix B

See Tables 7.7, 7.8, and 7.9.

Appendix C

See Tables 7.10, 7.11 and 7.12.

Table 7.7 Result of COE analysis—robust test: excluding 20 biggest firms from the sample

	ESG	*Environmental*			*Social*				*Governance*		
	ESG score	*Resource use score*	*Emissionsscore*	*Environmental innovation score*	*Workforce score*	*Human rights score*	*Community score*	*Product responsibility score*	*Management score*	*Shareholders score*	*CSR strategy score*
	A.1	*A.2*	*A.3*	*A.4*	*A.5*	*A.6*	*A.7*	*A.8*	*A.9*	*A.10*	*A.11*
Coefficient	**−0.1581***	−0.0150	−0.0114	−1.1533	**−0.0425****	−0.0089	0.0034	−0.0453	0.0134	−0.0536	**−0.0582****
	(0.0851)	(0.0243)	(0.0157)	(0.1333)	(0.0199)	(0.0869)	(0.0369)	(0.0328)	(0.0283)	(0.0526)	(0.0261)
Total assets	−0.0457	−0.3920	−0.4013	−0.6891	−0.3536	−0.4241	−0.4113	−0.4980	−0.3922	−0.4250	−0.3772
	(0.5133)	(0.5258)	(0.4997)	(0.7506)	(0.4954)	(0.5155)	(0.0524)	(0.5034)	(0.5059)	(0.5006)	(0.5018)
Total assets^2	0.0146	0.0123	0.0126	0.0219	0.0111	0.0133	0.0129	0.0157	0.0123	0.0133	0.0119
	(0.0164)	(0.0167)	(0.0159)	(0.0242)	(0.0158)	(0.0165)	(0.0167)	(0.0159)	(0.0160)	(0.0159)	(0.0160)
Long-term growth	0.0683	0.0698	0.0695	0.0652	0.0656	0.0696	0.0694	0.0685	0.0703	0.0693	0.0741
	(0.0425)	(0.0424)	(0.0424)	(0.0439)	(0.0428)	(0.0427)	(0.0424)	(0.0424)	(0.0429)	(0.0426)	(0.0434)
Market to book value	0.0495	0.0457	0.4631	0.0362	0.0501	0.0453	0.0455	0.0488	0.0450	0.0508	0.0474
	(0.0409)	(0.0404)	(0.0406)	(0.0419)	(0.0407)	(0.0403)	(0.0402)	(0.0414)	(0.0406)	(0.0405)	(0.0411)
No. observations	598	598	598	598	598	598	598	598	598	598	598

***, **, * stand for statistical significance at 1%, 5% and 10%, respectively

This table presents the results obtained by employing a half-logarithmic fixed effect regression model with clustered robust errors. Each column contains the output of regressions that considers Cost of Equity (COE) as dependent variable and the ESG scores as main independent variable. The clustered robust errors are presented in brackets. The control variables employed into the model are: Total assets; Long Term Growth, Market to Book Value; Time Trend. The employed models are in a half-logarithmic form: all the control variables and the main explanatory variable (ESG score) are in logarithmic form meanwhile the dependent variable (COE) is not

Source Thomson Reuters Datastream; Authors' elaboration

Table 7.8 Result of COE analysis—robust test: excluding 20 smallest firms from the sample

	ESG	*Environmental*			*Social*				*Governance*		
	ESG score	*Resource use score*	*Emissions score*	*Environmental innovation score*	*Workforce score*	*Human rights score*	*Community score*	*Product responsibility score*	*Management score*	*Shareholders score*	*CSR strategy score*
	B.1	*B.2*	*B.3*	*B.4*	*B.5*	*B.6*	*B.7*	*B.8*	*B.9*	*B.10*	*B.11*
Coefficient	**−0.1668**** (0.0811)	−0.0162 (0.0208)	−0.0080 (0.0149)	−0.0779 (0.0963)	**−0.0436**** (0.0208)	0.0341 (0.0807)	0.0107 (0.0391)	**−0.0664*** (0.0343)	−0.0062 (0.0271)	−0.0448 (0.0433)	**−0.0601**** (0.0254)
Total assets	−0.7791 (0.6458)	−0.7544 (0.6576)	−0.7709 (0.6401)	−0.9568 (0.7369)	−0.7310 (0.6421)	−0.7583 (0.6417)	−0.7842 (0.6497)	−0.8810 (0.6544)	−0.7960 (0.6485)	−0.7218 (0.6154)	−0.7430 (0.6416)
Total assets^2	0.2433 (0.0202)	0.0234 (0.0206)	0.0239 (0.0201)	0.0297 (0.0231)	0.0226 (0.0201)	0.0234 (0.0201)	0.0243 (0.0203)	0.0273 (0.0204)	0.0247 (0.0203)	0.0222 (0.0193)	0.0231 (0.0201)
Long-term growth	0.0584 (0.0368)	0.0600 (0.0366)	0.0598 (0.0366)	0.0574 (0.0374)	0.0556 (0.0371)	0.0592 (0.0371)	0.0594 (0.0365)	0.0590 (0.0366)	0.0596 (0.0369)	0.0614 (0.0362)	0.0631 (0.0371)
Market to book value	0.0477 (0.0442)	0.0455 (0.0442)	0.0462 (0.0441)	0.0403 (0.0452)	0.0497 (0.0441)	0.0476 (0.0440)	0.0462 (0.0442)	0.0494 (0.0449)	0.0463 (0.0442)	0.0497 (0.0446)	0.0476 (0.0443)
No. observations	657	657	657	657	657	657	657	657	657	657	657

***, **, * stand for statistical significance at 1%, 5% and 10%, respectively

This table presents the results obtained by employing a half-logarithmic fixed effect regression model with clustered robust errors. Each column contains the output of regressions that considers Cost of Equity (COE) as dependent variable and the ESG scores as main independent variable. The clustered robust errors are presented in brackets. The control variables employed into the model are: Total assets; Long Term Growth, Market to Book Value; Time Trend. The employed models are in a half-logarithmic form: all the control variables and the main explanatory variable (ESG score) are in logarithmic form meanwhile the dependent variable (COE) is not

Source Thomson Reuters Datastream; Authors' elaboration

Table 7.9 Result of COE analysis—robust test: period 2010–2018

	ESG	*Environmental*			*Social*				*Governance*		
	ESG score	*Resource use score*	*Emissions score*	*Environmental innovation score*	*Workforce score*	*Human rights score*	*Community score*	*Product responsibility score*	*Management score*	*Shareholders score*	*CSR strategy score*
	C.1	*C.2*	*C.3*	*C.4*	*C.5*	*C.6*	*C.7*	*C.8*	*C.9*	*C.10*	*C.11*
Coefficient	**−0.3694***	−0.0667	−0.0431	−0.1816	−0.0618	0.0936	0.0044	**−0.1724****	0.0330	−0.0933	**−0.1411****
	(0.1983)	(0.0432)	(0.0542)	(0.1573)	(0.0483)	(0.1631)	(0.0967)	(0.0775)	(0.0536)	(0.0819)	(0.0627)
Total assets	−0.1798	−1.9206	−1.9392	−2.1564	−1.9105	−2.0656	−2.0138	−2.0355	−2.0580	−1.9075	−1.8575
	(1.2035)	(1.2293)	(1.2114)	(1.3646)	(1.4892)	(1.1935)	(1.2191)	(1.1683)	(1.2164)	(1.2547)	(1.2946)
Total assets^2	0.0585	0.0621	0.0627	0.0700	0.0617	0.0661	0.0649	0.0666	0.0662	0.0614	0.0602
	(0.0392)	(0.0400)	(0.0394)	(0.0449)	(0.0388)	(0.0390)	(0.0396)	(0.0386)	(0.0396)	(0.0377)	(0.0416)
Long term growth	0.0854	0.0888	0.0894	0.0860	0.0846	0.0883	0.0894	0.0932	0.0906	0.0937	0.0975
	(0.0462)	(0.0469)	(0.0456)	(0.0467)	(0.0450)	(0.0463)	(0.0459)	(0.0457)	(0.0459)	(0.0451)	(0.0463)
Market to book value	0.2076	0.1964	0.1988	0.1847	0.2005	0.1960	0.1977	0.2248	0.1963	0.2096	0.2008
	(0.0962)	(0.0959)	(0.0959)	(0.0985)	(0.0961)	(0.0971)	(0.0960)	(0.0986)	(0.0963)	(0.0969)	(0.0959)
No. observations	504	504	504	504	504	504	504	504	504	504	504

***, **, * stand for statistical significance at 1%, 5% and 10%, respectively

This table presents the results obtained by employing a half-logarithmic fixed effect regression model with clustered robust errors. Each column contains the output of regressions that considers Cost of Equity (COE) as dependent variable and the ESG scores as main independent variable. The clustered robust errors are presented in brackets. The control variables employed into the model are: Total assets; Long Term Growth, Market to Book Value; Time Trend. The employed models are in a half-logarithmic form: all the control variables and the main explanatory variable (ESG score) are in logarithmic form meanwhile the dependent variable (COE) is not

Source Thomson Reuters Datastream; Authors' elaboration

Table 7.10 Result of FP analysis—robust test: excluding 20 biggest firms from the sample

	ESG	*Environmental*			*Social*				*Governance*		
	ESG score	*Resource use score*	*Emissions score*	*Environmental innovation score*	*Workforce score*	*Human rights score*	*Community score*	*Product responsibility score*	*Management score*	*Shareholders score*	*CSR strategy score*
	D.1	*D.2*	*D.3*	*D.4*	*D.5*	*D.6*	*D.7*	*D.8*	*D.9*	*D.10*	*D.11*
Coefficient	**−0.0429***** (0.0134)	**−0.0103***** (0.0040)	−0.0057 (0.0045)	−0.0009 (0.0044)	**−0.0057*** (0.0031)	−0.0102 (0.0100)	**−0.0104***** (0.0035)	−0.0046 (0.0053)	−0.0039 (0.0054)	**−0.0061**** (0.0029)	**−0.0076**** (0.0036)
Total assets	−0.0941 (0.0832)	−0.0934 (0.0812)	−0.0921 (0.0804)	−0.0932 (0.0805)	−0.0879 (0.0819)	−0.0987 (0.0828)	−0.0947 (0.0829)	−0.0969 (0.0819)	−0.0952 (0.0794)	−0.0920 (0.0803)	−0.0882 (0.0819)
Total assets^2	0.0028 (0.0026)	0.0027 (0.0026)	0.0026 (0.0026)	0.0026 (0.0025)	0.0025 (0.0026)	0.0028 (0.0026)	0.0027 (0.0026)	0.0027 (0.0026)	0.0027 (0.0025)	0.0025 (0.0025)	0.0025 (0.0026)
Market to book value	0.0372 (0.0084)	0.0368 (0.0084)	0.0373 (0.0085)	0.0370 (0.0085)	0.0372 (0.0084)	0.0367 (0.0085)	0.0372 (0.0085)	0.0373 (0.0084)	0.0373 (0.0083)	0.0373 (0.0084)	0.0368 (0.0084)
Leverage	−0.0325 (0.0071)	−0.0322 (0.0071)	−0.0326 (0.0072)	−0.0325 (0.0072)	−0.0326 (0.0072)	−0.0324 (0.0072)	−0.0323 (0.0072)	−0.0326 (0.0072)	−0.0327 (0.0072)	−0.0327 (0.0072)	−0.0325 (0.0072)
ROA (*t* − 1)	0.1990 (0.0446)	0.2118 (0.4395)	0.2154 (0.0436)	0.2172 (0.0434)	0.2128 (0.0430)	0.2153 (0.0431)	0.2115 (0.0437)	0.2152 (0.0441)	0.2129 (0.0439)	0.2146 (0.0436)	0.2135 (0.0439)
No. observations	1,693	1,693	1,693	1,693	1,693	1,693	1,693	1,693	1,693	1,693	1,693

***, **, * stand for statistical significance at 1%, 5% and 10%, respectively
This table presents the results obtained by employing a half-logarithmic fixed effect regression model with clustered robust errors. Each column contains the output of regressions that considers Return on Assets (ROA) as dependent variable and the ESG scores as main independent variable. The clustered robust errors are presented in brackets. The control variables employed into the model are: Total assets; Financial Leverage, Market to Book Value; Past Profitability (ROA lagged one period). The employed models are in a half-logarithmic form: all the control variables and the main explanatory variable (ESG score) are in logarithmic form meanwhile the dependent variable (ROA) and the past profitability measure are not
Source Thomson Reuters Datastream; Authors' elaboration

Table 7.11 Result of FP analysis—robust test: excluding 20 smallest firms from the sample

	ESG	*Environmental*			*Social*				*Governance*		
	ESG Score	*Resource use score*	*Emissions score*	*Environmental innovation score*	*Work-force score*	*Human rights score*	*Community score*	*Product responsi-bility score*	*Management score*	*Share-holders score*	*CSR strategy score*
	E.1	*E.2*	*E.3*	*E.4*	*E.5*	*E.6*	*E.7*	*E.8*	*E.9*	*E.10*	*E.11*
Coefficient	**−0.0323***** (0.0114)	**−0.0007**** (0.0031)	−0.0035 (0.0036)	0.0013 (0.0017)	−0.00233 (0.0027)	−0.0128 (0.0081)	**−0.0070***** (0.0027)	−0.0011 (0.0047)	**−0.0077***** (0.0026)	−0.0034 (0.0021)	**−0.0093***** (0.0035)
Total assets	0.0120 (0.0420)	0.0056 (0.0401)	0.0048 (0.0396)	0.0039 (0.0391)	0.0063 (0.0397)	−0.0021 (0.0417)	0.0096 (0.0413)	0.0024 (0.0405)	0.0058 (0.0391)	0.0083 (0.0396)	0.0052 (0.0418)
Total assets^2	−0.0008 (0.0013)	−0.0007 (0.0012)	−0.0007 (0.0012)	−0.0007 (0.0012)	−0.0007 (0.0012)	−0.0004 (0.0013)	−0.0008 (0.0013)	−0.0006 (0.0012)	−0.0007 (0.0012)	−0.0008 (0.0012)	−0.0007 (0.0013)
Market to book value	0.0239 (0.0069)	0.0236 (0.0069)	0.0237 (0.0069)	0.0236 (0.0069)	0.0237 (0.0069)	0.0232 (0.0069)	0.0235 (0.0070)	0.0236 (0.0068)	0.0239 (0.0068)	0.0237 (0.0069)	0.0233 (0.0069)
Leverage	−0.0290 (0.0071)	−0.0289 (0.0071)	−0.0290 (0.0072)	−0.0291 (0.0072)	−0.0290 (0.0072)	−0.0289 (0.0071)	−0.0289 (0.0072)	−0.0290 (0.0716)	−0.0292 (0.0072)	−0.0291 (0.0072)	−0.0291 (0.0071)
ROA ($t-1$)	0.2219 (0.0523)	0.2317 (0.0513)	0.2339 (0.0511)	0.2356 (0.0507)	0.2334 (0.0505)	0.2354 (0.0512)	0.2321 (0.0508)	0.2353 (0.0513)	0.2321 (0.0512)	0.2342 (0.0508)	0.2305 (0.0503)
No. observa-tions	1808	1808	1808	1808	1808	1808	1808	1808	1808	1808	1808

***, **, * stand for statistical significance at 1%, 5% and 10%, respectively

This table presents the results obtained by employing a half-logarithmic fixed effect regression model with clustered robust errors. Each column contains the output of regressions that considers Return on Assets (ROA) as dependent variable and the ESG scores as main independent variable. The clustered robust errors are presented in brackets. The control variables employed into the model are: Total assets; Financial Leverage, Market to Book Value; Past Profitability (ROA lagged one period). The employed models are in a half- logarithmic form: all the control variables and the main explanatory variable (ESG score) are in logarithmic form meanwhile the dependent variable (ROA) and the past profitability measure are not

Source Thomson Reuters Datastream; Authors' elaboration

Table 7.12 Result of FP analysis—robust test: period 2010–2018

	ESG	*Environmental*			*Social*				*Governance*		
	ESG score	*Resource use score*	*Emission score*	*Environmental innovation score*	*Work-force score*	*Human rights score*	*Community score*	*Product responsi-bility score*	*Management score*	*Share-holders score*	*CSR strategy score*
	F.1	*F.2*	*F.3*	*F.4*	*F.5*	*F.6*	*F.7*	*F.8*	*F.9*	*F.10*	*F.11*
Coefficient	**−0.0531***** (0.0204)	**−0.0144**** (0.0064)	−0.0034 (0.0084)	−0.0012 (0.0053)	−0.0052 (0.0060)	−.0091 (0.0114)	**−0.0136**** (0.0063)	−0.0054 (0.0057)	−0.0038 (0.0094)	**−0.0062*** (0.0037)	**−0.0130*** (0.0055)
Total assets	−0.0425 (0.1214)	−0.0581 (0.1188)	−0.0558 (0.1157)	−0.0595 (0.1179)	−0.0532 (0.1195)	−0.0598 (0.1179)	−0.0625 (0.1203)	−0.0625 (0.1191)	−0.5450 (0.1239)	−0.0530 (0.1162)	−0.0522 (0.1195)
Total assets^2	0.0021 (0.0039)	0.0025 (0.0039)	0.0024 (0.0038)	0.0025 (0.0038)	0.0023 (0.0039)	0.0025 (0.0038)	0.0026 (0.0039)	0.0026 (0.0039)	0.0024 (0.0040)	0.0023 (0.0038)	0.0023 (0.0039)
Market to book value	0.0358 (0.0131)	0.0361 (0.0130)	0.0369 (0.0132)	0.0366 (0.0130)	0.0364 (0.0131)	0.0367 (0.0130)	0.0358 (0.0128)	0.0370 (0.0129)	0.0367 (0.0130)	0.0367 (0.0130)	0.0359 (0.0130)
Leverage	−0.0501 (0.0097)	−0.0503 (0.0097)	−0.0511 (0.0098)	−0.0512 (0.0099)	−0.0511 (0.0099)	−0.0510 (0.0098)	−0.0504 (0.0098)	−0.0512 (0.0098)	−0.0512 (0.0099)	−0.0514 (0.0098)	−0.0509 (0.0098)
ROA (t − 1)	0.0685 (0.0433)	0.0776 (0.0425)	0.0804 (0.0436)	0.0806 (0.0436)	0.0787 (0.0433)	0.0797 (0.0439)	0.0803 (0.0439)	0.0790 (0.0442)	0.0789 (0.0439)	0.0795 (0.0437)	0.0757 (0.0434)

	ESG	*Environmental*			*Social*				*Governance*		
	ESG score	*Resource use score*	*Emission score*	*Environmental innovation score*	*Work-force score*	*Human rights score*	*Community score*	*Product responsi-bility score*	*Management score*	*Share-holders score*	*CSR strategy score*
	F.1	*F.2*	*F.3*	*F.4*	*F.5*	*F.6*	*F.7*	*F.8*	*F.9*	*F.10*	*F.11*
No. observa-tions	1337	1337	1337	1337	1337	1337	1337	1337	1337	1337	1337

***, **, * stand for statistical significance at 1%, 5% and 10%, respectively

This table presents the results obtained by employing a half-logarithmic fixed effect regression model with clustered robust errors. Each column contains the output of regressions that considers Return on Assets (ROA) as dependent variable and the ESG scores as main independent variable. The clustered robust errors are presented in brackets. The control variables employed into the model are: Total assets; Financial Leverage, Market to Book Value; Past Profitability (ROA lagged one period). The employed models are in a half-logarithmic form: all the control variables and the main explanatory variable (ESG score) are in logarithmic form meanwhile the dependent variable (ROA) and the past profitability measure are not

Source Thomson Reuters Datastream; Authors' elaboration

References

Chen, K., Chen, Z., & Wei, K. (2009). Legal protection of investors, corporate governance, and the cost of equity capital. *Journal of Corporate Finance, 15*(3), 273–389.

Dhaliwal, D. S., Li, O. Z., Tsang, A., & Yang, Y. G. (2011). Voluntary nonfinancial disclosure and the cost of equity capital: The initiation of corporate social responsibility reporting. *The Accounting Review, 86*(1), 59–100.

Dhaliwal, D., Li, O. Z., Tsang, A., & Yang, Y. G. (2014). Corporate social responsibility disclosure and the cost of equity capital: The roles of stakeholder orientation and financial transparency. *Journal of Accounting Public Policy, 33*(4), 328–355.

Dowell, G., Hart, S., & Yeung, B. (2000). Do corporate global environmental standards create or destroy market value? *Management Science, 46*(8), 1013–1169.

Easton, P. D. (2004). PE ratios, PEG ratios, and estimating the implied expected rate of return on equity capital. *The accounting review, 79*(1), 73–95.

El Ghoul, S., Guedhami, O., Kim, H., & Park, K. (2018). Corporate environmental responsibility and the cost of capital: International evidence. *Journal of Business Ethics, 149*(2), 335–36126.

El Ghoul, S., Guedhami, O., Kwok, C. C. Y., & Mishra, D. R. (2011). Does corporate social responsibility affect the cost of capital? *Journal of Banking & Finance, 35*(9), 2388–2406.

Fama, E. F., & French, K. R. (1993). Common risk factors in the returns on stock and bonds. *Journal of Financial Economics, 33*, 3–56.

Filippetti, S. (2019). *I signori del lusso*. Sperling & Kupfer.

Gregory, A., Whittaker, J., & Yan, X. (2016). Corporate social performance, competitive advantage, earnings persistence and firm value. *Journal of Business Finance & Accounting, 43*(1), 3–30.

Gupta, K. (2018). Environmental sustainability and implied cost of equity: International evidence. *Journal of Business Ethics, 147*(2), 343–365.

Hail, L., & Leuz, C. (2011). *Cost of capital effects and changes in growth expectations around U.S. cross-listings* (ECGI - Finance Working Paper No. 46/2004, Wharton Financial Institutions Center Working Paper Series #06-19).

Hart, S. L., & Ahuja, G. (1996). Does it pay to be green? An empirical examination of the relationship between emission reduction and firm performance. *Business Strategy and the Environment, 5*(1), 30–37.

Konar, S., & Cohen, M. A. (2001). Does the market value environmental performance? *Review of Economics and Statistics, 83*(2), 281–289.

Lee, J., Graves, S. B., & Waddock, S. (2018). Doing good does not preclude doing well: Corporate responsibility and financial performance. *Social Responsibility Journal, 14*(4), 768–781.

Matthiesen, M.-L., & Salzmann, A. J. (2017). Corporate social responsibility and firms' cost of equity: How does culture matter? *Cross Cultural & Strategic Management, 24*(1), 105–124.

Reverte, C. (2012). The impact of better corporate social responsibility disclosure on the cost of equity capital. *Corporate Social Responsibility and Environmental Management, 19*(5), 253–272.

Sharfman, M. P., & Fernando, C. S. (2008). Environmental risk management and the cost of capital. *Strategic Management Journal, 29*(6), 569–592.

Suto, M., & Takehara, H. (2017). CSR and cost of capital: Evidence from Japan. *Social Responsibility Journal, 13*(4), 798–816.

Tamburi Investment Partners. (2017). *Prezzi & valori. L'enterprise value nell'era digitale. Borsa, private equity, M&A, premi, sconti, errori e prospettive*. Class Editori.

Waddock, S. (2017). Inequality, dignity and the sustainability challenge. *Journal of Management for Global Sustainability, 5*(1), 63–83.

CHAPTER 8

Automotive and Tire Sector: Sustainability as the New "Value" for Shareholders

Carlo Bellavite Pellegrini, Raul Caruso, and Rocco Cifone

Introduction

Although corporate finance has historically researched about the determinants of stock returns and modeling future yields, recently the corporate governance has focused its attention on measuring the impact of

C. Bellavite Pellegrini
Department of Economic Policy, Catholic University of the Sacred Heart, Milan, Italy
e-mail: carlo.bellavite@unicatt.it

R. Caruso (✉)
Research Centre of Applied Economics, Catholic University of the Sacred Heart, Milan, Italy
e-mail: raul.caruso@unicatt.it

R. Cifone
Catholic University of the Sacred Heart, Milan, Italy

© The Author(s), under exclusive license to Springer Nature Switzerland AG 2022

C. B. Pellegrini et al. (eds.), *Climate Change Adaptation, Governance and New Issues of Value*, Palgrave Studies in Impact Finance,
https://doi.org/10.1007/978-3-030-90115-8_8

non-financial information on listed companies' corporate financial performance. This field of study has become more relevant over time due to the increasing attention of investors. In economic literature, the search for a relation between Environmental, Social and Governance scores (ESG) and corporate financial performance can be traced back to the beginning of the 1970s.[1]

A large growing literature is nowadays investigating to what extent sustainable strategies affect both firm's performance and value. The challenge of this empirical approach is to verify whether considering sustainability, environmental and social issues also payoff in terms of performance and added value to the firm. Whether it is reasonable to say that such strategies of firms do contribute to the establishment of a more sustainable business context as envisioned in Waddock (2017), there are still substantial doubts about the role of ESG strategies in shaping both profitability and firm value [see among others Lee et al. (2018)]. In particular, the ESG scores combine elements which separately have already proven to affect firm performance and value of firms.[2] In sum, albeit fragmented there is already a large literature on the topic. Friede et al. (2015) found more than 2,000 studies that analyze the ESG and financial performance link.

As noted above, therefore, this chapter addresses the relationship between firm profitability and firm's value and ESG in the automotive and tire sector over the period between 2002 and 2016 for a panel of European and North American listed companies. In fact, when we focus on listed companies, it is always important to remember that there is a strong difference between price and value (Tamburi Investment Partners, 2017).

Focusing on a specific sector is worthwhile for several reasons. First and foremost, in such a way we rule out the risk of distortion in the results plausibly driven by heterogeneity between sectors. Secondly, when focusing on a specific sector it is easier to target also a recurring doubt in the existing literature, namely the direction of causality between environmental and social behavior and firm performance. In fact, firms with

[1] See Chapter 6.

[2] In fact, for instance, there is a copious literature on the relationship between firm performance and environmental issues [see among others Dowell et al. (2000), Hart and Ahuja (1996), Konar and Cohen (2001)].

better performance can be capable of investing more resources in environmental and social strategies, so generating a self-reinforcing positive association between ESG (or its component) and profitability [see on this point Waddock and Graves (1998)].

The dataset collects 48 listed firms from Europe and North America. The total sample is composed of 11 firms that produce automobiles and 37 that produce components. In the first analysis, the dependent variable is the Return on Asset (ROA), computed as the ratio between EBITDA and Total Assets. In the second analysis, the dependent variable is Tobin's Q, which is computed as the sum of Market Value of Equity and the Book Value of Assets reduced by the Book Value of Equity and by the Deferred Taxes, all divided by the Book Value of Assets.

On both analyses we have employed a parsimonious set of control variables drawn from the established literature: a size variable, a leverage variable and an efficiency variable captured by means of Total Asset Turnover. Another section is devoted to the analysis of the relationship between ESG and profitability in the tire sector.

Our results show that the ESG score is positively associated with firms' profitability captured by means of Returns on Assets (ROA). In brief, when the ESG score increases by 10% the profitability measure increases by 0.04. Yet, among the components of overall ESG, the environmental score is the one that exhibits an association in a linear model so suggesting that overall results are mainly driven by this. When the environmental score increases by 10% the profitability measure increases by 0.014. Eventually when considering the firm value proxied by means of Tobin's Q, results also show a negative association between Tobin's Q and both the environmental component of ESG. In particular, a 10% increase in the one-year lagged environmental score translates into a current reduction in Tobin's Q of −0.01.

Eventually further estimations have highlighted a more nuanced evidence. First, we have considered the interaction between the ESG scores and the firm size (captured by means of total assets). Findings show that as the firm size increases the relationship between both environmental and social components of ESG and firm's profitability turns to be negative. Instead, no significant interaction emerges when considering the Tobin's Q as dependent variable. Yet, there is an inverse U-shaped relationship between the governance score of ESG and ROA of firms. Eventually we have considered non-linearities. Results show that when governance score is small, ROA of firms slightly decreases but as the governance scores increases it eventually increases. In other words, there is an inverse U-shaped relationship between the governance score and the firm's profitability.

Data and the Empirical Strategy

The ESG scores are taken from the dataset Thomson Reuters Datastream. The data provider captures and calculates over 400 company-level ESG measures, of which they select a subset of 178 most comparable and relevant fields to power the overall company assessment and scoring process.[3] The underlying measures are based on considerations around comparability, data availability and industry relevance. They are grouped into ten categories, weighted proportionately to the count of measures within each category formulates the final ESG Score, which reflects the company's ESG performance, commitment and effectiveness based on publicly reported information. The current study will employ all ten categories, the overall Environmental, Social and Governance scores and the synthetic ESG score.

The dataset collects 48 listed firms from Europe and North America. The total sample is composed of 11 firms that produce automobiles (22.92% of total sample) and 37 that produce components (77.08% of total sample). In 2017 the worldwide number of listed firms exhibiting ESG scores in the automotive sector was 131.

Figure 8.1 shows the total market capitalization at the end of each year and the number of listed companies that compose the sample. The trend of market capitalization shows a moderate growth between 2002 and 2007. After the crisis, the capitalization recovered from its previous loss and showed a sustained growth, peaking in 2015 and flexing slightly thereafter. Table 8.1 shows the geographical distribution of the sample.

In order to increase the robustness of the analysis, a set of control variables is employed and they are all drawn from Thomson Reuters Datastream. Because of data availability, our data spans from 2002 to 2016. Table 8.2 reports the descriptive statistics of the variables used in the regression.

[3] The fact that the data provider already modifies the ESG criteria according to the sector it belongs to, strengthens the analysis approach of this study which considers a single sector.

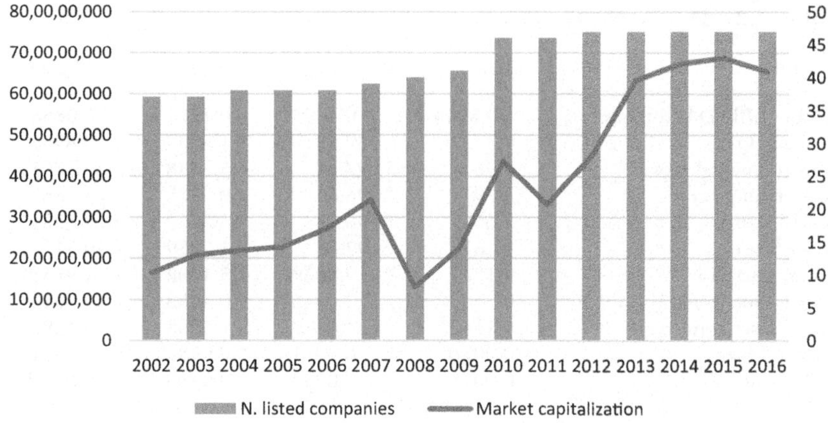

Fig. 8.1 Total capitalization (thousands of Euros) and number of firms (2002–2016) (*Source* Thomson Reuters Datastream, authors' elaboration)

Table 8.1 Number of firms

Headquarter nations	*No. automobiles companies*	*No. auto components companies*	*Total companies*	*Percentage on total sample*
Canada	0	3	3	6.25
Finland	0	1	1	2.08
France	2	4	6	12.50
Germany	4	4	8	16.67
Ireland	0	1	1	2.08
Italy	0	1	1	2.08
Sweden	0	1	1	2.08
United Kingdom	0	1	1	2.08
UK-Netherland	1	0	1	2.08
United States	4	21	25	52.08
Total	**11**	**37**	**48**	**100.00**

Source Thomson Reuters Datastream, authors' elaboration

ESG Scores and ROA

In order to analyze the relationship between ESG scores and firm profitability, we employ the following baseline OLS panel fixed effects

Table 8.2 Descriptive statistics

Variable	*Min*	*Max*	*Mean*	*Std. deviation*
ROA (EBITDA/total assets)	−0.5944	0.4892	0.1231	0.0858
Tobin's Q	0.5722	9.53	1.5357	0.8956
Environmental score	8.95	97.48	73.3188	27.2503
Governance score	2.79	96.61	60.6832	24.2865
Social score	5.65	98.75	65.3452	28.5531
ESG Score	11.57	92.86	51.7959	19.3476
Total asset (Ln)	9.1247	19.8069	15.4302	1.9583
Total debt/enterprise value	0	2.0946	0.4463	0.2619
Total asset turnover	0.1642	2.6449	1.1868	0.4295
Capex/asset	0.0026	0.2871	0.0582	0.0351
R&D/sales	0.0018	1.0232	0.0408	0.0662

Source Thomson Reuters Refinitiv, authors' elaboration

model:

$$\ln \mathrm{ROA}_{it} = \alpha + \beta_1 \ln \mathrm{ESG}_{it} + \beta_2 \ln X_{it} + \beta_5 \mathrm{ROA}_{it-1} + \tau + \varepsilon_{it}$$

The dependent variable is the Return on Asset (ROA), computed as the ratio between Earnings Before Interest, Tax and Depreciation (EBITDA) and total asset.[4] We also employ a parsimonious set of control variables established in the existing literature: a size variable, represented by the natural logarithm of total assets, a leverage variable, represented by the natural logarithm of the ratio between total debt and an accounting measure of enterprise value, an efficiency variable captured by means of the Total Asset Turnover, that is the ratio between net sales and revenues and total asset. Yet, we also include the one-year lagged profitability since these values are commonly highly correlated with past values. In the light of the results of the Hausman test,[5] we employ a fixed-effects model. Year dummies are also included. With the exception of the lagged dependent

[4] The above definition of ROA may be compared to a normalization of company's EBITDA on a measure of accounting size of the same company. In this regard EBITDA may be seen as the most similar accounting measure to the notion of unlevered cash flows.

[5] Hausman test identifies which model is the most appropriate between the fixed-effects model and the random-effects model.

Table 8.3 Return on Asset and ESG scores

Dependent variable ROA defined as EBITDA/total asset; FE model								
	8.3.1	*8.3.2*	*8.3.3*	*8.3.4*	*8.3.5*	*8.3.6*	*8.3.7*	*8.3.8*
Log of environmental score	**0.018*** (**0.0101**)	**0.015**** (**0.007**)						
Log of governance score			0.001 (0.0111)	0.004 (0.010)				
Log of social score					0.010 (0.011)	0.012 (0.010)		
Log of ESG score							**0.037*** (**0.022**)	**0.041**** (**0.019**)
Control variables	Yes	Yes	Yes	Yes	Yes	Yes	Yes	Yes
Lagged dependent variable	No	Yes***	No	Yes***	No	Yes***	No	Yes***
Year dummies	Yes	Yes	Yes	Yes	Yes	Yes	Yes	Yes
Constant	−0.552 (0.514)	−0.226 (0.437)	−0.497 (0.520)	−0.179 (0.452)	−0.530 (0.513)	−0.2112 (0.432)	−0.5921 (0.519)	−0.274 (0.429)
Observations	387	365	387	365	387	365	380	360
Number of firms	47	47	47	47	47	47	47	47
R-squared:								
Within	0.359	0.4331	0.3541	0.4297	0.3565	0.4326	0.3688	0.4465
Between	0.0207	0.2216	0.0197	0.2154	0.0191	0.2125	0.0144	0.2031
Overall	0.0461	0.2424	0.0449	0.237	0.043	0.2323	0.0374	0.2324

Robust errors in brackets; errors are clustered on firm; ***, ** and * denotes significance at the 1%, 5% and 10% level, respectively

Source Thomson Reuters Datastream-Refinitiv, authors' elaboration

variable, all predictor variables are log-transformed by means of natural logarithm.

Table 8.3 reports the results of the regression between ROA and ESG scores.[6] The overall ESG score does exhibits a positive association with ROA. In particular, when including the one-year lagged value of ROA, the association between ESG and current ROA gets more robust so reaching a 5% threshold of statistical significance. Control variables do

[6] For sake of readability coefficients of control variables are not displayed.

exhibit the expected signs so not suggesting concerns about the general fitness of the model. In order to compute the quantitative effect we can say that for a 10% increase in ESG score, the difference in the expected mean ROA value is: 0.41*ln (1.1) = 0.04. In brief, when the ESG score increases by 10% the profitability measure increases by 0.04. Since values of ROA in our samples are bounded between −0.59 and 0.49 (as it is shown in Table 8.2), this result is by no means trivial. In particular, this result appears to be mainly driven by the environmental component of ESG. In fact, only the latter appears to be significantly associated with profitability. When the environmental score increases by 10% the profitability measure increases by 0.014. This first regression shows that investing in ESG, especially the environmental one, increases the profitability in terms of ROA of companies belonging to the automotive and parts sector.

In the light of the results, it is reasonable to investigate further on the environmental components. Eventually we have run the baseline regression by including simultaneously the three distinct factors of the environmental score. Results are shown in Table 8.4. The capacity of a firm to reduce the use of resources finding more eco-efficient solutions appears to be negatively associated with the profitability measure. The plausible interpretation is that costs associated with such processes may have a negative impact on profitability. However, the impact seems to be negligible. The negative quantitative impact on ROA of a 10% increase in the resource use score is −0.001.

Instead, the emission reduction score is positively associated with firms' profitability and a 0.01 increase in ROA measure would be associated with a 10% increase in company's commitment in reducing environmental emissions.

Eventually we have estimated the interaction of ESG variables and the firm size. Once added an interaction term between the firm size (captured by means of the log of total assets) and the different ESG metrics, the results deliver a more nuanced evidence. First, the social score of the ESG gains statistical significance and it is positively associated with firms' profitability. In particular, in model 8.5.6 including the lagged measure of ROA, there would be a 0.02 increase in the profitability measure associated with a 10% increase in the social score. The interaction terms between firm size and both environmental score and social score are negative, so suggesting that as the firm size increases the different ESG scores are negatively associated with profitability. The plausible interpretation

Table 8.4 Environmental components of ESG and profitability

Dependent variable ROA defined as EBITDA/total asset; FE model		
	8.4.1	*8.4.2*
Log of resource use score	**−0.008****	−0.004
	(0.005)	(0.003)
Log of emissions score	**0.012***	0.005
	(0.007)	(0.006)
Log of environmental innovation score	0.012	0.008
	(0.008)	(0.007)
Control variables	Yes	Yes
Lagged dependent variable	No	Yes
Year dummies	Yes	Yes
Constant	−0.495	−0.194
	(0.487)	(0.438)
Observations	371	351
Number of firms	43	43
R-squared:		
Within	0.3736	0.4373
Between	0.0065	0.1725
Overall	0.0373	0.2242

Standard errors in brackets; standard errors are clustered on firm; ***, ** and * denotes significance at the 1%, 5% and 10% level, respectively

Source Thomson Reuters Datastream-Refinitiv, authors' elaboration

in broad terms is that as the firm size grows the range of costs associated with the commitment to environmental and social issues prove to become detrimental for firms' profitability[7]. In simpler words, it is likely that adaptation costs are increasing in the size of firms (Table 8.5).

Tobin's Q and ESG Scores

Eventually we also estimate an OLS regression using Tobin's Q as the dependent variable and ESG score as main explanatory variables. Tobin's Q is commonly used as proxy for the firm value or for firm's perspectives of profitability as it is intended to capture the value of long-term investments including intangibles. In brief, it is often used as proxy of

[7] We have to remember however that size is not necessarily a proxy for an increasing company's capitalization.

Table 8.5 Interactions between ESG scores and firm size

Dependent variable ROA as EBITDA/total asset; FE model								
	8.5.1	*8.5.2*	*8.5.3*	*8.5.4*	*8.5.5*	*8.5.6*	*8.5.7*	*8.5.8*
Log of environmental score	**0.397**** **(0.209)**	**0.270*** **(0.165)**						
Log of governance score			−0.017 (0.092)	0.012 (0.076)				
Log of social score					**0.276**** **(0.125)**	**0.220**** **(0.111)**		
Log of ESG score							**0.257*** **(0.155)**	0.200 (0.153)
Log of firm size (total asset)	**0.139**** **(0.070)**	0.088 (0.056)	0.036 (0.027)	0.021 (0.024)	**0.097**** **(0.047)**	0.066 (0.043)	**0.089*** **(0.051)**	0.054 (0.049)
Log of firm size * log of environmental score	**−0.025**** **(0.014)**	−0.017 (0.011)						
Log of firm size * log of governance score			0.001 (0.006)	−0.001 (0.005)				
Log of firm size * log of Social score					**−0.017**** **(0.008)**	**−0.014**** **(0.007)**		
Log of firm size * log of ESG score							−0.014 (0.009)	−0.010 (0.010)
Other control variables	Yes	Yes	Yes	Yes	Yes	Yes	Yes	Yes
Lagged dependent variable	No	Yes***	No	Yes***	No	Yes***	No	Yes***
Year dummies	Yes	Yes	Yes	Yes	Yes	Yes	Yes	Yes
Constant	−2.067** (1.059)	−1.266 (0.855)	−0.432 (0.443)	−0.210 (0.395)	**−1.391**** **(0.729)**	−0.916 (0.670)	−1.378 (0.817)	−0.867 (0.768)
Observations	387	365	387	365	387	365	380	360

(continued)

Table 8.5 (continued)

Dependent variable ROA as EBITDA/total asset; FE model								
	8.5.1	*8.5.2*	*8.5.3*	*8.5.4*	*8.5.5*	*8.5.6*	*8.5.7*	*8.5.8*
Number of firms	47	47	47	47	47	47	47	47
R-squared:								
Within	0.3754	0.4412	0.3543	0.4297	0.3765	0.4426	0.3759	0.4505
Between	0.0435	0.2387	0.0178	0.2199	0.0756	0.2946	0.0279	0.2349
Overall	0.1011	0.2824	0.0428	0.2399	0.1302	0.3097	0.0632	0.2673

Standard errors in brackets; standard errors are clustered on firm; ***, ** and * denotes significance at the 1%, 5% and 10% level, respectively
Source Thomson Reuters Datastream-Refinitiv, authors' elaboration

firm's value. In our context, Fernando et al. (2016), for example, find that the Tobin's Q appears to be negatively correlated with environmental performance of firms. We employ the following definition:

$$\text{Tobin's Q} = \frac{\text{market value of equity} + \text{book value of asset} - \text{book value of equity} - \text{balance sheet deferred taxes}}{\text{book value of asset}}$$

Then, we employ a slightly different model from the previous one. Firstly, in line with the existing literature, all predictor variables are one-year lagged. This determines a reduction in number of observations. In fact, profitability has an impact on firm value and so we include also here the one-year lagged ROA in the regression model as predictor variable. The control variables we employ are: the ratio between R&D and sales, the ratio between the total debt and a measure of enterprise value and the ratio between capex and assets. All explanatory variables are log-transformed by means of natural logarithm. In notations the empirical model to estimate determinant of Tobin's Q is:

$$\ln(\text{Tobin's Q})_{it} = \alpha_0 + \alpha_1 \ln(\text{ESG})_{it-1} + Z_{it-1} + \tau + \varepsilon_{it}$$

where $(\text{ESG})_{it-1}$ denotes alternatively the overall ESG score and its determinants, Z_{it-1} is the set of control variables and τ a year dummy. Errors are clustered on firm. In this case we also show results of a fixed-effects model. The regressions estimate shows that only the environmental components of ESG exhibit a significant statistical association with the Tobin's Q. Such relationship is negative. The impact is not negligible. In particular, a 10% increase in the one-year lagged environmental score translates into a current reduction in Tobin's Q of −0.01. The plausible interpretation of the negative relationship is that environmental commitment increases the replacement costs of the firm so finally decreasing the Tobin's Q. In this respect, size of firms presumably matters (Table 8.6).

Table 8.6 Tobin's Q and ESG scores

Dependent variable Tobin's Q; FE model								
	8.6.1	*8.6.2*	*8.6.3*	*8.6.4*	*8.6.5*	*8.6.6*	*8.6.7*	*8.6.8*
Log of environmental score ($t-1$)	**−0.094**** (0.045)	**−0.097**** (0.042)						
Log of governance score ($t-1$)			0.028 (0.045)	0.034 (0.046)				
Log of social score ($t-1$)					0.002 (0.054)	0.001 (0.053)		
Log of ESG score							−0.007 (0.087)	−0.019 (0.0913)
Control variables ($t-1$)	Yes	Yes	Yes	Yes	Yes	Yes	Yes	Yes
Years dummies	Yes	Yes	Yes	Yes	Yes	Yes	Yes	Yes
Profitability ($t-1$)	No	Yes***	No	Yes***	No	Yes***	No	Yes***
Constant	3.713*** (0.596)	3.703*** (0.683)	3.506*** (0.533)	3.500*** (0.6159	3.411*** (0.742)	3.399*** (0.861)	3.373*** (0.638)	3.392*** (0.741)
Observations	290	287	290	287	290	287	283	280
Number of firms	35	35	35	35	35	35	35	35
R-squared:								
Within	0.4666	0.478	0.4588	0.4709	0.4563	0.4672	0.4511	0.4616
Between	0.4513	0.4439	0.4407	0.4306	0.4445	0.4343	0.4614	0.4525
Overall	0.476	0.4624	0.4576	0.442	0.4614	0.4462	0.4805	0.4673

Standard errors in brackets; standard errors are clustered on firm; ***, ** and * denotes significance at the 1%, 5% and 10% level, respectively

Source Thomson Reuters Datastream-Refinitiv, authors' elaboration

Table 8.7 Environmental components of ESG and Tobin's Q

Dependent variable Tobin's Q; FE model		
	8.7.1	*8.7.2*
Log of resource use score ($t-1$)	−0.002	0.006
	(0.022)	(0.024)
Log of emissions score ($t-1$)	−0.003	−0.011
	(0.025)	(0.027)
Log of environmental innovation score ($t-1$)	0.023	0.018
	(0.030)	(0.031)
Control variables ($t-1$)	Yes	Yes
Year dummies	Yes	Yes
Profitability ($t-1$)	No	Yes
Constant	3.332***	3.309***
	(0.554)	(0.627)
Observations	279	276
Number of companies	34	34
R-squared:		
Within	0.4573	0.4675
Between	0.4585	0.4454
Overall	0.4755	0.4597

Standard errors in brackets; standard errors are clustered on firm; ***, ** and * denotes significance at the 1%, 5% and 10% level, respectively
Source Thomson Reuters Datastream-Refinitiv, authors' elaboration

Eventually also for the Tobin's Q we have estimated the different components of the environmental score. However, no significant results take shape (see Table 8.7). Yet, we have introduced in the baseline regression an interaction of ESG variables and the firm size (see Table 8.8). Also, in this case we do not find significant results for the ESG factors. Interestingly, the insight envisioned in Table 8.6 seems to be confirmed because also in Table 8.8 the size of firms (captured by means of the log of total asset) is negatively associated with Tobin's Q.

Non-linearities

In light of the previous results, we employ a simple robustness test seeking for a non-linear relationship between the ESG scores and the dependent variables. Results are puzzled and suggest the existence of non-linearities in the relationship we are investigating. Then, we add to the regressions the quadratic term of the ESG score and its components. Results

Table 8.8 Interactions between ESG scores and firm size

Dependent variable Tobin's Q; FE model

	8.8.1	*8.8.2*	*8.8.3*	*8.8.4*	*8.8.5*	*8.8.6*	*8.8.7*	*8.8.8*
Log of environmental score ($t-1$)	−0.467 (0.563)	−0.940 (0.652)						
Log of governance score ($t-1$)			0.105 (0.316)	0.174 (0.324)				
Log of social score ($t-1$)					−0.245 (0.435)	−0.486 (0.570)		
Log of ESG Score ($t-1$)							−0.397 (0.582)	−0.566 (0.595)
Log of Firm Size (total asset) ($t-1$)	**−0.260*** **(0.144)**	**−0.397**** **(0.183)**	**−0.164***** **(0.066)**	**−0.162***** **(0.067)**	**−0.225**** **(0.109)**	**−0.289**** **(0.150)**	**−0.250*** **(0.149)**	**−0.291**** **(0.158)**
Log of firm size * log of environmental score ($t-1$)	0.0242 (0.0363)	0.055 (0.042)						
Log of firm size * log of governance score ($t-1$)			−0.005 (0.0194)	−0.009 (0.020)				
Log of firm size * log of Social score ($t-1$)					0.016 (0.026)	0.031 (0.035)		
Log of firm size * log of ESG score ($t-1$)							0.024 (0.037)	0.033 (0.038)
Control variables ($t-1$)	Yes	Yes	Yes	Yes	Yes	Yes	Yes	Yes
Year fixed effects	Yes	Yes	Yes	Yes	Yes	Yes	Yes	Yes
Profitability ($t-1$)	No	Yes***	No	Yes***	No	Yes***	No	Yes***
Constant	5.317*** (2.205)	7.253*** (2.786)	3.261*** (1.147)	3.056*** (1.147)	4.303*** (1.734)	5.152** (2.336)	4.823** (2.315)	5.419** (2.44)
Observations	290	287	290	287	290	287	283	280
Number of companies	35	35	35	35	35	35	35	35

Dependent variable Tobin's Q; FE model

	8.8.1	*8.8.2*	*8.8.3*	*8.8.4*	*8.8.5*	*8.8.6*	*8.8.7*	*8.8.8*
R-squared:								
Within	0.4676	0.4827	0.4591	0.4719	0.458	0.473	0.453	0.4653
Between	0.4405	0.4201	0.4368	0.424	0.4476	0.4391	0.4576	0.4473
Overall	0.4713	0.4524	0.454	0.4358	0.4676	0.4589	0.4817	0.4695

Standard errors in brackets; standard errors are clustered on firm; ***, ** and * denotes significance at the 1%, 5% and 10% level, respectively
Source Thomson Reuters Datastream-Refinitiv, authors' elaboration

are puzzled and deliver a more nuanced evidence (Table 8.9 reports the results).

Differently from regressions previously shown, the governance score appears to be more relevant in determining profitability of firms. In particular, there is an inverse U-shaped relationship between the governance score and current ROA. The coefficient of the quadratic term is positive suggesting that when the governance score is small, profitability of firms decreases. Eventually as the governance scores increases profitability of firms increases too, namely the relationship appears to become positive even if the coefficient is rather small (Bellavite Pellegrini et al., 2011, 2017).

The overall ESG turns to be statistically insignificant. Eventually, also the relationship between Tobin's Q appears to be characterized by non-linearities as shown in Table 8.10. First, it seems there is an inverse U-shaped relationship between environmental scores and Tobin's Q, since the coefficient of the quadratic term is positive. In fact, this would mean that when the ESG score is small the Tobin's Q decreases. Eventually, as the environmental score increases the Tobin's Q increases. This result appears to be robust. The plausible interpretation of such result is that a minor commitment to environmental issues turns to be detrimental for firm value because of the costs associated, but eventually as it becomes higher the firm value increase because of the efficiency gains due to advancements in technology and efficiency.

Insights About the Relationship Between Profitability and ESG in the Tire Market

Among the firms considered, in particular seven out of 48 are tire producers, namely Michelin, Continental, Nokian, Pirelli, Cooper Tire & Rubber, Goodyear Tire & Rubber.[8] Because of the scarcity of data, we have to proceed using a different methodology. Differently from the previous sections, we employ two measures of profitability, namely EBITDA and ROA.[9] Therefore, since we are not able to identify a punctual causation, rather we can identify the existence of stable relationships between our measures of profitability and ESG factors. First, we compute

[8] Titan has been excluded because of lack of data.

[9] Computed also in this case as the ratio between EBITDA and total asset.

Table 8.9 Return on Asset and ESG scores—non-linearities

Dependent variable ROA defined as EBITDA/total asset; FE model								
	8.9.1	*8.9.2*	*8.9.3*	*8.9.4*	*8.9.5*	*8.9.6*	*8.9.7*	*8.9.8*
Log of environmental score	−0.008 (0.067)	−0.039 (0.064)						
Log of environmental score squared	0.005 (0.010)	0.007 (0.009)						
Log of governance score			**−0.144**** **(0.066)**	**−0.122**** **(0.56)**				
Log of governance score squared			**0.021***** **(0.009)**	**0.018***** **(0.008)**				
Log of social score					−0.068 (0.060)	−0.038 (0.049)		
Log of social score squared					0.011 (0.009)	0.007 (0.007)		
Log of ESG score							0.160 (0.165)	0.170 (0.154)
Log of ESG score squared							−0.016 (0.0221)	−0.0174 (0.021)
Control variables	Yes	Yes	Yes	Yes	Yes	Yes	Yes	Yes
Lagged dependent variable	No	Yes***	No	Yes***	No	Yes***	No	Yes***
Year dummies	Yes	Yes	Yes	Yes	Yes	Yes	Yes	Yes
Constant	−0.506 (0.540)	−0.133 (0.457)	−0.394 (0.409)	−0.128 (0.362)	−0.410 (0.533)	−0.142 (0.426)	−0.806 (0.571)	−0.513 (0.504)
Observations	387	365	387	365	387	365	380	360
Number of firms	47	47	47	47	47	47	47	47
R-squared:								
Within	0.3592	0.4339	0.3727	0.4443	0.3598	0.434	0.3706	0.4487
Between	0.021	0.2196	0.0094	0.1533	0.0115	0.1936	0.0184	0.2251

(continued)

Table 8.9 (continued)

Dependent variable ROA defined as EBITDA/total asset; FE model								
	8.9.1	*8.9.2*	*8.9.3*	*8.9.4*	*8.9.5*	*8.9.6*	*8.9.7*	*8.9.8*
Overall	0.0462	0.24	0.0271	0.1779	0.0335	0.2169	0.044	0.2527

Standard errors in brackets; standard errors are clustered on firm; ***, ** and * denotes significance at the 1%, 5% and 10% level, respectively

Source Thomson Reuters Datastream-Refinitiv, authors' elaboration

the Pearson correlation coefficients: the correlation between ROA and ESG exhibits a small coefficient (0.05), whereas the latter presents a correlation of 0.66. Eventually, in order to have some insights on the relationship between ESG scores and firms' profitability in the tire sector, we first analyze some plots reporting both ROA and EBITDA against the ESG score (Figs. 8.2 and 8.3). From a visual analysis, it seems that ROA—albeit weakly—is positively associated with the aggregate ESG score. Instead when considering the EBITDA, such positive association appears to be more compelling. Figure 8.3 shows that high ESG score are associated with high EBITDA values (Table 8.11).

When considering the environmental component of ESG, the correlation coefficients are 0.23 and 0.30, respectively, for ROA and EBITDA. Eventually we present the plot on the relationships between the different components of ESG and both ROA and EBITDA. Also, in this case, the line of fitted values in most cases has an upward slope (Figs. 8.4, 8.5, 8.6, 8.7, 8.8 and 8.9).

Besides correlation, eventually we have run the Johansen cointegration test to check whether it is possible to identify a stable relationship between the variables under investigation. The results show that—with only one exception—there is cointegration between our profitability measures (ROA and EBITDA) and all the ESG scores in the long-run. In brief, the cointegration tests confirm that there is a relationship between profitability and ESG scores in the long-run.

Finally, we have run the test for Granger causality. The Granger causality test is usually adopted to understand whether one variable does include enough information to predict another variable. The intuition behind the Granger causality is that a variable can be predicted not only by its own past value but also by past values of another variable. That is, albeit informative, Granger causality is not to be interpreted as a true causality,

Table 8.10 Tobin's Q and ESG scores—non-linearities

Dependent variable Tobin's Q; FE model								
	8.10.1	*8.10.2*	*8.10.3*	*8.10.4*	*8.10.5*	*8.10.6*	*8.10.7*	*8.10.8*
Log of environmental score ($t-1$)	**−0.943*** **(0.520)**	**−1.048**** **(0.523)**						
Log of environmental score ($t-1$) squared	0.111 (0.070)	**0.125*** **(0.070)**						
Log of governance score ($t-1$)			0.021 (0.224)	0.0770 (0.232)				
Log of governance score squared			0.001 (0.031)	−0.006 (0.032)				
Log of social score ($t-1$)					−0.495 (0.835)	−0.420 (0.841)		
Log of social score squared					0.067 (0.109)	0.0569 (0.110)		
Log of ESG Score							0.165 (0.612)	0.132 (0.607)
Log of ESG score squared							−0.022 (0.083)	−0.020 (0.082)
Control variables ($t-1$)	Yes	Yes	Yes	Yes	Yes	Yes	Yes	Yes
Year dummies	Yes	Yes	Yes	Yes	Yes	Yes	Yes	Yes
Profitability ($t-1$)	No	Yes***	No	Yes***	No	Yes*	No	Yes***
Constant	5.378*** (0.958)	5.542*** (1.064)	3.507*** (0.548)	3.494 (0.629)	4.425** (2.118)	4.260 (2.314)	3.052 (1.225)	3.109 (1.382)
Observations	290	287	290	287	290	287	283	280
Number of firms	35	35	35	35	35	35	35	35

(continued)

Table 8.10 (continued)

Dependent variable Tobin's Q; FE model								
	8.10.1	*8.10.2*	*8.10.3*	*8.10.4*	*8.10.5*	*8.10.6*	*8.10.7*	*8.10.8*
R-squared:								
Within	0.4727	0.4856	0.4588	0.471	0.5633	0.4721	0.4512	0.4617
Between	0.399	0.3887	0.4414	0.4268	0.47	0.4595	0.4661	0.4562
Overall	0.4416	0.4259	0.4581	0.4394	0.4818	0.4662	0.4828	0.469

Standard errors in brackets; standard errors are clustered on firm; ***, ** and * denotes significance at the 1%, 5% and 10% level, respectively

Source Thomson Reuters Datastream-Refinitiv, authors' elaboration

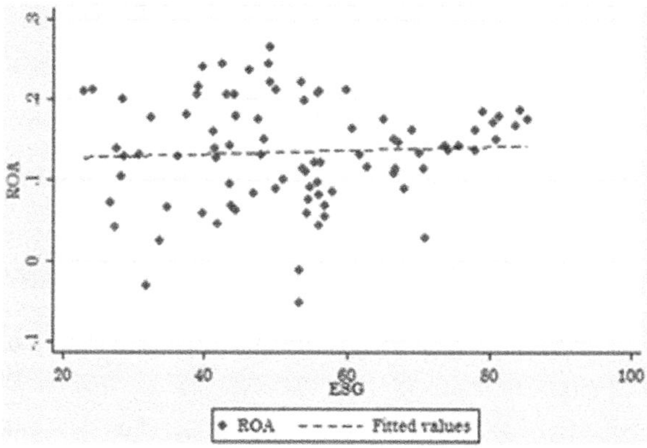

Fig. 8.2 ROA and ESG (2002–2016)

but rather an additional explanation to determine whether one time-series can be considered useful to predict another one. Because of the short time-span the Dumitrescu and Hurlin (2012) test is to be applied. Then, in our context we first tested whether the aggregate ESG score does Granger-cause ROA and EBITDA finding a negative result. ESG does not Granger-cause both ROA and EBITDA. When considering the environmental score, findings show that it does Granger-cause EBITDA but not ROA. Instead the governance score does Granger-cause ROA but not EBITDA. Lastly, the social score does Granger-cause both ROA and EBITDA. In simpler words, interpreting the results in the light of the proper meaning of Granger causality, we can maintain that EBITDA

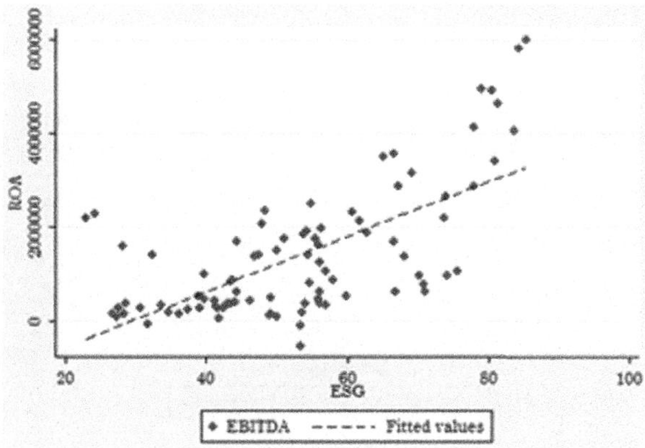

Fig. 8.3 EBITDA and ESG (2002–2016)

Table 8.11 Pearson correlation coefficients

	Environmental	*Corporate*	*Social*	*ESG*
EBITDA	0.30	0.06	0.49	0.66
ROA (EBITDA/total asset)	0.23	−0.21	−0.07	0.05

Source Thomson Reuters Datastream-Refinitiv, authors' elaboration

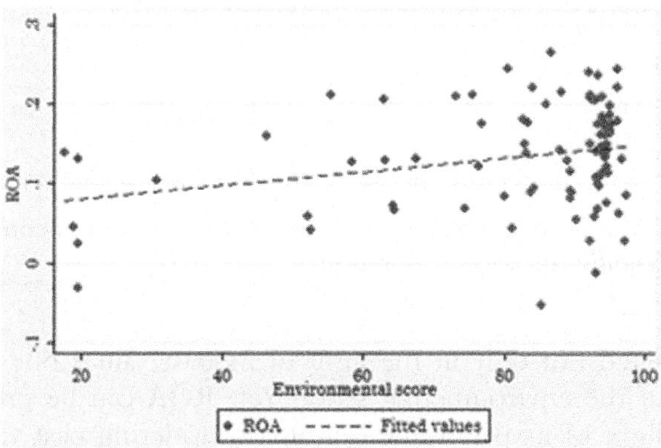

Fig. 8.4 ROA and environmental score (2002–2016) (*Source* Thomson Reuters Datastream-Refinitiv, authors' elaboration)

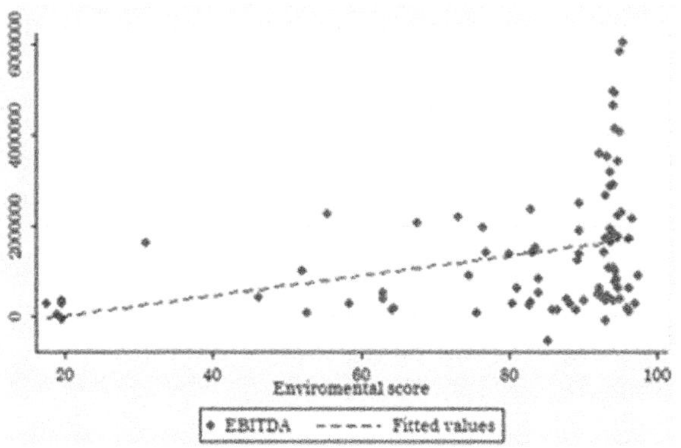

Fig. 8.5 EBITDA and Environmental score (2002–2016) (*Source* Thomson Reuters Datastream-Refinitiv, authors' elaboration)

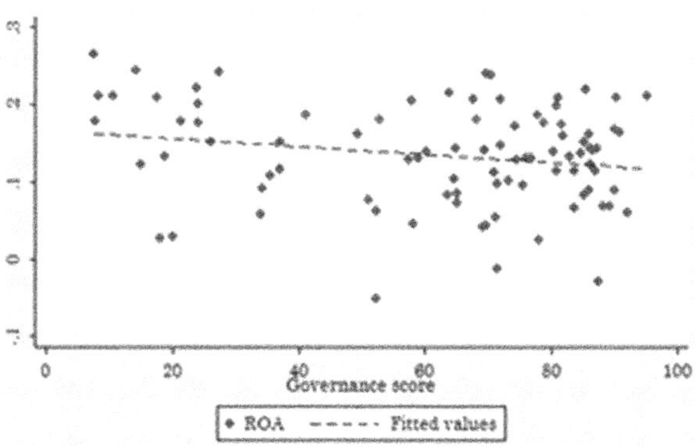

Fig. 8.6 ROA and governance score (2002–2016) (*Source* Thomson Reuters Datastream-Refinitiv, authors' elaboration)

can be predicted not only in the light of its past values but also on the past values of the environmental score. Yet, ROA can be predicted not only in the light of its past value but also considering past values of the social score. In sum, this result contributes to support the idea that also in the tire market profitability is associated with some components of ESG. Table 8.12 summarizes the Granger causality results.

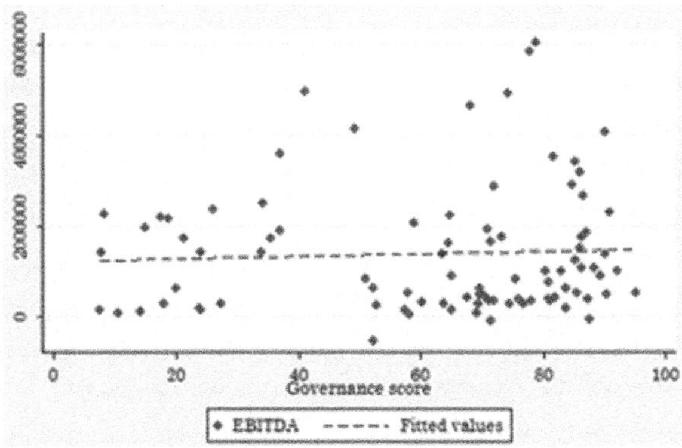

Fig. 8.7 EBITDA and governance score (2002–2016) (*Source* Thomson Reuters Datastream-Refinitiv, authors' elaboration)

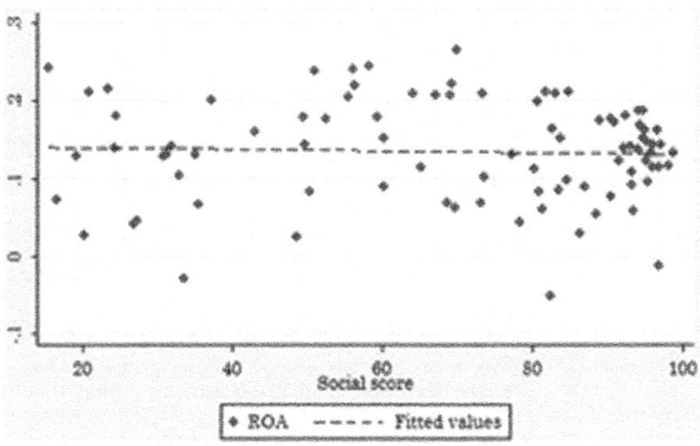

Fig. 8.8 ROA and social score (2002–2016) (*Source* Thomson Reuters Datastream-Refinitiv, authors' elaboration)

In sum, albeit cautiously, it can be maintained that there is a relationship between ESG factors and profitability for tire producers in North America and Europe. In particular, it seems that the commitment to environmental improvement does payoff by pushing the EBITDA up. This result holds for the social score too whereas improvements in the governance factor translate into a more stable relationship with ROA.

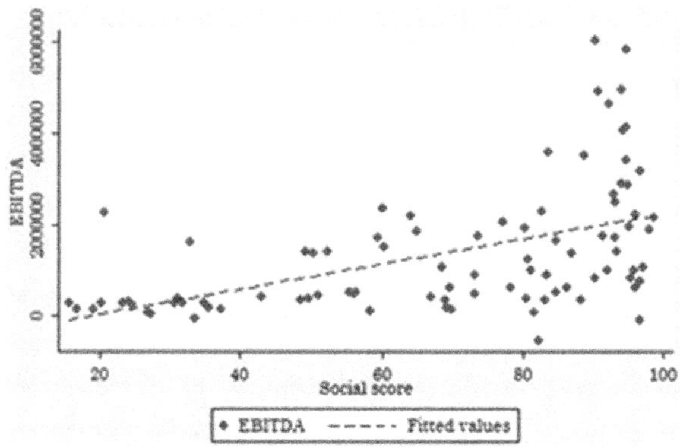

Fig. 8.9 EBITDA and social score (2002–2016) (*Source* Thomson Reuters Datastream-Refinitiv, authors' elaboration)

Table 8.12 Granger causality

ESG	ROA	No
	EBITDA	No
Environmental score	ROA	No
	EBITDA	Yes
Governance score	ROA	Yes
	EBITDA	No
Social score	ROA	Yes
	EBITDA	Yes

Results based on Dumitrescu and Hurlin (2012) Granger causality test
Source Thomson Reuters Datastream-Refinitiv, authors' elaboration

Summary of the Results and Discussion

This work focused on the impact of ESG scores on profitability of a panel of North-American and European listed firms in the automotive and parts sector over the period between 2002 and 2016. In sum our main findings highlight that:

i. the aggregate ESG score is positively associated with ROA[10] of firms. When the ESG score increases by 10% the profitability

[10] Computed as the ration between EBITDA and total assets.

measure increases by 0.04. Please note that values ROA in our sample are bounded between −0.59 and 0.49. In brief the result is by no means trivial.

ii. among the components of the aggregate ESG, the environmental score is the one that exhibits an association in a linear model so suggesting that overall results are mainly driven by this. When the environmental score increases by 10% the profitability measure increases by 0.014.

iii. there is a negative association between the Tobin's Q and the environmental component of ESG. In particular, a 10% increase in the one-year lagged environmental score translates into a current reduction in Tobin's Q of −0.01.

These main findings are to be complemented with some additional results which emerged when studying interactions between different variables and non-linearities. In particular we found that:

iv. there is an inverse U-shaped relationship between the governance score and ROA, namely when the governance score is small, ROA of firms decreases. As the governance scores increases, ROA increases.

v. the size of firms matters because when controlling for the interaction term between the different ESG components and the size of firms it emerges that: (v.i) the interaction between the environmental score and the size of firm is negatively associated with ROA and (v.ii) the interaction between the social score and the size of firm is negatively associated with ROA. In simpler words, in both cases, it is likely that adaptation costs are increasing in the size of firms.

In brief, there is a nuanced evidence on the impact of ESG scores on profitability of firms. In particular, from the methodological point of view, it ought to be noted that any proper analysis on this topic cannot rely on aggregate ESG scores only. Albeit informative, the ESG aggregate measure needs to be split into its components in order to derive more properly usable insights for managers and investors. Needless to say, the main limitation of this work descends from the lack of data. Development of ESG scores are a very recent advancement and therefore data availability is small.

In particular, since the time-series is not long enough to evaluate properly the temporal effect of ESG aggregate score and its components is not analyzed in depth. In other words, we still have a little understanding of time horizon over which a better performance in ESG scores translates into an improved profitability. Further research would extend the dataset including Asian tires and automotive companies and would also consider second-level subcomponents, namely, the subcomponents of environmental, social and governance scores, respectively.

Appendix A

See Table 8.13.

Table 8.13 Definition of variables

Variable	*Definition*	*Source*
ROA_{it}	It is the ratio between Earnings Before Interest Taxes and Depreciation (EBITDA) and Total Asset	Thomson Reuters Datastream
Tobin Q	It is the natural logarithm of the ratio between the market value of equity plus the book value of asset minus the book value of equity and the deferred taxes all on book value of asset	Thomson Reuters Datastream
$Environmental_{it}$	The environmental pillar measures a company's impact on living and non-living natural systems, including the air, land and water, as well as complete ecosystems. It reflects how well a company uses best management practices to avoid environmental risks and capitalize on environmental opportunities in order to generate long-term shareholder value	Thomson Reuters Datastream

(continued)

Table 8.13 (continued)

Variable	*Definition*	*Source*
$Social_{it}$	The social pillar measures a company's capacity to generate trust and loyalty with its workforce, customers and society, through its use of best management practices. It is a reflection of the company's reputation and the health of its license to operate, which are key factors in determining its ability to generate long-term shareholder value	Thomson Reuters Datastream
$Governance_{it}$	The corporate governance pillar measures a company's systems and processes, which ensure that its board members and executives act in the best interests of its long-term shareholders. It reflects a company's capacity, through its use of best management practices, to direct and control its rights and responsibilities through the creation of incentives, as well as checks and balances in order to generate long-term shareholder value	Thomson Reuters Datastream
ESG $Score_{it}$	ESG Score is an overall company score based on the self-reported information in the environmental, social and corporate governance pillars	Thomson Reuters Datastream
Total $Asset_{it}$	It is the natural logarithm of total asset	Thomson Reuters Datastream
Total debt/Enterprise Value	It is the ratio between total debt to a measure of enterprise value. [The Enterprise value is defined as (i) common equity + (ii) Preferred stock + (iii) Minority Interest + (iv) Long Term Debt + (v) Short term debt and current portion of long term debt]	Thomson Reuters Datastream
Total asset turn $over_{it}$	It is the ratio between net sales and revenue and total asset	Thomson Reuters Datastream
Capex/$Asset_{it}$	It is the ratio between Capex and total asset	Thomson Reuters Datastream

Source Authors' elaboration

Appendix B

See Table 8.14.

Table 8.14 Companies included in the sample

Company name	*Headquarter*	*Stock market*	*No. years obs*
American Axle & Manufacturing	USA	USA-NYSE	15
Aptiv	Ireland	USA-NYSE	5
Autoliv	Sweden	USA-NYSE-Sweden	15
BMW	Germany	Germany	15
BorgWarner	USA	USA-NYSE	15
Continental	Germany	Germany	15
Cooper Tire & Rubber Company	USA	USA-NYSE	15
Cooper-Standard Automotive	USA	USA-NYSE	7
Daimler	Germany	Germany	15
Dana	USA	USA-NYSE	9
Dorman Products	USA	USA-NASDAQ	15
ElringKlinger	Germany	Germany	15
Faurecia	France	France	15
Fiat Chrysler Automobiles	UK—TheNetherlands	USA-NYSE	15
Ford Motor Company	USA	USA-NYSE	15
General Motors Company	USA	USA-NYSE	7
Gentex Corporation	USA	USA-NYSE	15
Gentherm Incorporated	USA	USA-NYSE	15
Genuine Parts Company	USA	USA-NYSE	15
GKN	UK	UK	15
Goodyear Tire & Rubber Company	USA	USA-NYSE	15
Groupe Renault	France	France	15
Harley-Davidson	USA	USA-NYSE	15
Lear Corporation	USA	USA-NYSE	8
Leoni	Germany	Germany	15
Linamar Corporation	Canada	Canada	15

(continued)

Table 8.14 (continued)

Company name	*Headquarter*	*Stock market*	*No. years obs*
LKQ corporation	USA	USA-NASDAQ	13
Magna international	Canada	USA-NYSE and Canada	15
Martinrea international	Canada	Canada	15
Michelin	France	France	15
Modine Manufacturing	USA	USA-NYSE	15
Motorcar Parts of America	USA	USA-NASDAQ	15
Nokian Tyres	Finland	Finland	15
Peugeot	France	France	15
Pirelli & C	Italy	Italy	13
Plastic Omnium	France	France	15
PORSCHE	Germany	Germany	15
Schaeffler Technologies	Germany	Germany	2
Standard Motor Products	USA	USA-NYSE	15
Stoneridge	USA	USA-NYSE	15
Tenneco	USA	USA-NYSE	15
Tesla	USA	USA-NASDAQ	7
Titan Tire Corporation	USA	USA-NASDAQ	15
Tower International	USA	USA-NYSE	7
Valeo	France	France	15
Visteon Corporation	USA	USA-NYSE	7
Volkswagen	Germany	Germany	15
WABCO holdings	USA	USA-NYSE	10

Source Authors' elaboration

References

Bellavite Pellegrini, C., Romelli, D., & Sironi, E. (2011). The impact of governance and productivity on stock returns in European industrial companies. *Investment Management and Financial Innovation, 8*(4), 20–28.

Bellavite Pellegrini, C., Sergi, B., & Sironi, E. (2017). Stock returns, productivity and corruption in eight European fast emerging markets. *Thunderbird International Business Review, 59*(1), 15–22.

Dowell, G., Hart, S., & Yeung, B. (2000). Do corporate global environmental standards create or destroy market value? *Management Science, 46*(8), 1013–1169.

Dumitrescu, E., & Hurlin, C. (2012). Testing for Granger non-causality in heterogenous panels. *Economic Modelling, 29*(4), 1450–1460.

Fernando, C. S., Sharfman, M. P., & Uysal, V. B. (2016). Corporate environmental policy and shareholder value: Following the smart money. *Journal of Financial and Quantitative Analysis, 52*(5), 2023–2051.

Friede, G., Busch, T., & Bassen, A. (2015). ESG and financial performance: Aggregated evidence from more than 2000 empirical studies. *Journal of Sustainable Finance & Investment, 5*(4), 210–233.

Hart, S. L., & Ahuja, G. (1996). Does it pay to be green? An empirical examination of the relationship between emission reduction and firm performance. *Business Strategy and the Environment, 5*(1), 30–37.

Konar, S., & Cohen, M. A. (2001). Does the market value environmental performance? *Review of Economics and Statistics, 83*(2), 281–289.

Lee, J., Graves, S. B., & Waddock, S. (2018). Doing good does not preclude doing well: Corporate responsibility and financial performance. *Social Responsibility Journal, 14*(4), 768–781.

Tamburi Investment Partners. (2017). *Prezzi & valori. L'enterprise value nell'era digitale. Borsa, private equity, M&A, premi, sconti, errori e prospettive.* Class Editori.

Waddock, S. (2017). Inequality, dignity and the sustainability challenge. *Journal of Management for Global Sustainability, 5*(1), 63–83.

Waddock, S., & Graves, S. B. (1998). The corporate social performance-financial link. *Strategic Management Journal, 18*(4), 303–319.

CHAPTER 9

How Much Does Matter ESG Ratings in Big Pharma Firms Performances?

Sophia Veronica Barbieri and Laura Pellegrini

Introduction

Nowadays, the pharmaceutical companies' contribution to sustainable development is one of the key concerns of investors and an increasing number of stakeholders (employees, consumers, firms, governments,

Sophia Veronica Barbieri wrote the following sections: Literature Reviews, The Pharmaceutical Sector Around the World: First Empirical Evidences, The Empirical Approaches: Highlights and results (Section ROA and ESG Scores—Q1–Q2).
Laura Pellegrini wrote the following sections: Introduction, The Empirical Approaches: Highlights and Results (Section TOBIN'S Q and ESG Scores—Q1–Q2) and Conclusions.

S. V. Barbieri (✉)
Milan, Italy

L. Pellegrini
Department of Economics and Finance, University of Bari "Aldo Moro", Bari, Italy
e-mail: laura.pellegrini@uniba.it

© The Author(s), under exclusive license to Springer Nature Switzerland AG 2022

C. B. Pellegrini et al. (eds.), *Climate Change Adaptation, Governance and New Issues of Value*, Palgrave Studies in Impact Finance,
https://doi.org/10.1007/978-3-030-90115-8_9

media, investors). The pharmaceutical sector is one of the main sectors that invests most in Research & Development, in order to develop new therapies and launch a substantial number of new products on the market. Since it should be an industry that aims to improve public health, its stakeholders are very attentive to sustainable policies adopted by pharmaceutical companies. However, part of the social acceptance of this firms is compromised due to the constant corporate scandals (corruption, bribes, price manipulation, promotion for unauthorized use, fraud to the health system), mainly because of rules infringements from employees, and controversial quality and safety of products being sold.[1] So, following this increased attention from stakeholder, it becomes essential for companies to invest resources in sustainable strategies if they do not want to lose potential significant capitals.

In literature, it is possible to find several studies that have tried to highlight the existence of a positive relationship between good sustainability standards and corporate performance. However, without reaching unambiguous and consistent results. Existing researches provide consistent evidence of a predominantly positive relationship between good sustainability standards and firm performance (Freeman, 1984), but also negative (Friedman, 1970) or neutral (Ullmann, 1985). From this overview, emerges the key question that, in this chapter focused on empirical issues, it is tried to answer: is there really a positive relationship between ESG indicators and business success, measured in terms of performance? To answer this question, using an econometric model for panel data, it is verified the existence of a relationship between the firm performance and different ESG scores: the ESG Score, the scores of its three pillars and their sub-components, the ESG Controversies Score and the ESG Combined Score. The controversies index is also included, as a measure of the impact of media news on corporate performance because media news has a significant influence on the companies' financial behavior.

Therefore, this chapter aims to analyze whether it is convenient for pharmaceutical companies to invest resources in ESG strategies, in terms of economic and financial performance. According to these considerations, we analyze a sample of 103 listed companies around the world

[1] Many of the major global pharmaceutical companies were guilty of these violations (such as GlaxoSmithKline, Johnson & Johnson, Novartis, Pfizer). Their continued increase in sanctions and settlement agreements with authorities, along with remedial actions cost, undermines performance and social acceptance (Takatsuki, 2018).

that exhibit ESG scores between 2004 and 2018. Consistently with ESG researches, we approach to firm's profitability issues in terms of Return on Assets (ROA) and then to Tobin's Q, as a proxy of firm value in the long term.

At first, results highlight the existence of a statistically significant relationship between the score of the Governance component and the firm performance, showing a negative relation with ROA and a positive one with Tobin's Q. In particular, a 10% increase in the Governance score leads to a 21 bp reduction in ROA and an increase in Tobin's Q at time $t + 1$ of 0.70% (*ceteris paribus*). This result seems to be mainly driven by the score of the Management component. When the Management score increases by 10%, the firm profitability measure decreases by 16 bp and the firm value at time $t + 1$ increases by 0.46% (*ceteris paribus*). Furthermore, if we consider an interaction term between the firm size and ESG scores, all the variables on which research hypotheses were tested gain consistency. In particular, as the firm size increases, there is a negative relationship between the firm profitability and the Environment score; positive with Governance and Controversies scores. Instead, considering the firm value measure, as the firm size increases, there is a negative relationship with the score of ESG, Social, Governance and ESG Combined components; positive with the Controversies score. In quantitative terms, a 10% increase in the ESG Controversies Score leads to a reduction in ROA of 224 bp and a reduction in Tobin's Q at time $t + 1$ of 13.3% (*ceteris paribus*).

Finally, we point out an inverted U-shaped nonlinear relationship between the score of ESG, Social, Governance and ESG Combined components and Tobin's Q. Results of this inverse relationship suggest that when these scores are small, the firm value increases but, at some point, as the ESG scores increase, the firm value decreases.

Literature Reviews

Literature refers to the broad class of investment practices that integrate ESG factors with different names (e.g., Socially Responsible Investment, Sustainable Investment). Although there are some differences, Corporate Social Responsibility can generally be understood as approximately equivalent to ESG indicators, even though the latter are more related to the current sustainability concept. Since its origins, the Corporate

Social Responsibility theory has provided a compelling stakeholder-oriented alternative (Freeman, 1984) to the classic shareholder-centered view about maximizing the economic capital current value (Friedman, 1970), which involves environmental, social and governance factors. For example, growing public attention on top management remuneration, companies' bankruptcy and sustainable investment volumes, underlines that there is no mutual exclusion between contributing to shareholder and stakeholder value (Kitzmueller & Shimshack, 2012).

Various empirical studies confirm a positive relationship between sustainability and business performance. There is a common idea that companies that address environmental (Dowell et al., 2000), social (Cavaco & Crifo, 2014) and governance (Ferrell et al., 2016) risks have the ability to manage the overall company risk better than their competitors, thus presenting better performance. Social responsibility reduces corporate costs, creates value for stakeholders and internal capabilities (Preston & O'bannon, 1997). It also helps to generate positive relationships with customers, attracting motivated employees and reducing business risks, thus strengthening its reputation (Bird et al., 2007). All of this contributes to the firm competitive advantage, making it first in a specific sector. Therefore, by investing in superior social responsibility, a company creates a reputational capital stock and thus improves its financial performance. Furthermore, Derwall (2007) shows that companies that are perceived as more sustainable in the long term, can benefit from higher firm value. Since these types of firms are less risky, they enjoy higher firm value as investors demand lower returns. Similarly, Kim and Kim (2014) show that companies with high ESG scores benefit from higher Tobin's Q, due to the lower risk that characterizes these companies.

On the other side, many studies show that ESG controversies are negatively and directly linked to business performance. Increased media attention to corporate controversies triggers greater stakeholders' skepticism toward the company (Du et al., 2010), thus leading it a credibility and firm performance reduction (Godrey et al., 2009). Some of the conflicting results in literature derive from the assumption that positive and negative news share homogeneous costs and returns but have a reverse impact on firm value (Cho et al., 2013). Furthermore, a positive sustainable behavior is able to mitigate the effects of a negative one (Vanhamme & Grobben, 2009). As literature suggests, it would seem that higher ESG score is not only highly relevant to a better company performance but could help to mitigate negative effects due to sudden bad news.

Although most studies highlight positive links between good sustainability standards and economic-financial performance, others suggest the existence of negative or neutral relationships. In support of the negative relationship, Milton Friedman's current of thought (1970) claims that companies involved in social responsibility activities show a greater competitive weakness because they incur costs that other institutions would have had to bear (ecological operations, philanthropy, customer welfare, health centers, environmental protection). The only company goal should be to maximize profit, always respecting the legal and ethical decorum. Similarly, Hemingway and Maclagan (2004) believe that the focus on sustainable activities serves as a cover for management's fraudulent activities, showing a negative relationship with corporate performance. Skeptics accuse social responsibility of only serving to spread a good image of the company, regardless of its unethical practices that have not been publicized (Caulkin, 2002).

The debate on CSR and financial performance has led to another possibility, that this issue is independent and doesn't lead to any financial results, on the basis of a costs–benefits balance. The variable measuring social performance and the other one measuring financial performance are mutually exclusive and each relationship is merely coincidental. The exponent of this current of thought argues that there are so many variables that interpose themselves between CSR and financial performance that the relationship hardly exists (Ullmann, 1985). McWilliams and Siegel (2000) showed that any relationship between social responsibility and financial performance may depend on an incorrect model specification. In fact, after incorporating R&D expenses into the model, it was found that CSR has a neutral effect on performance. Further studies show that there is an indirect relationship between sustainable commitment and firm value, that depends on the company reputation about its socially responsible behavior toward stakeholders (Surroca et al., 2010). Finally, it was found that mainly the well-known companies (i.e. the larger and generally more visible ones) are able to benefit from higher firm value, due to an intense advertising on their socially responsible behavior, capable of influencing stakeholder (Servaes & Tamayo, 2013).

The Pharmaceutical Sector Around the World: First Empirical Evidences

In the highlight of these evidences and given the characteristics of the Pharmaceutical sector according to ESG features, our focus or research is to analyze the impact of ESG scores on firm performance, on a sample of 103 pharmaceutical listed companies around the world (21 countries)[2] that display ESG scores in the period between 2004 and 2018.[3] All data in the analysis are taken from the dataset Thomson Reuters Datastream.[4]

Most of the listed companies included in the sample are represented by US companies (36.89%), followed by Chinese (17.48%) and Hong Kong companies (8.74%); representing about 63% of the sample in terms of company's number. Instead, companies with the greatest weight on the total capitalization are US companies (51.77%), followed by Swiss (15.45%) and British companies (7.80%); which represent about 75% of the total capitalization.

According to the first evidences we can remark that more than half of the sample countries (Belgium, Brazil, Finland, France, Hungary, Indonesia, Israel, Italy, Japan, Cayman and Slovenia) do not record significant differences between the percentage weight on listed companies' total number and the weight on companies' total market capitalization. Instead, Canada, China, Spain, Hong Kong and South Korea have a percentage weight on companies' total number greater than the weight of their capitalization. On the other hand, Switzerland, Germany, Denmark, Great Britain and USA have a lower weight on the total number of listed companies compared to the percentage weight on company's total capitalization.

[2] Mainly, the sample is made-up of three geographical areas: North America, Asia and Europe.

[3] Only listed companies that close the financial statements on 31 December were considered. The company's total number exhibiting ESG scores on 12/31/2018 would be 111, but 8 of these companies were excluded from the sample, given that Thomson Reuters Datastream does not present some accounting and current data necessary for the empirical analysis.

[4] Given the recent focus on ESG issues, Thomson Reuters Datastream has no available data for the pharmaceutical sector before 2004. So, the 15 years observation period was chosen to incorporate the largest and most reliable ESG scores dataset, also considering the peculiarities of this specific sector.

Considering the sample distribution based on companies' size, measured by market capitalization we can underline that most of the companies included in the sample (33.0%) belongs to the capitalization range between 1 and 5 billion of US$. Followed by companies with a market capitalization of less than 1 billion (around 19.4%) and companies with a capitalization of between 5 and 10 billion of US$ (around 18.4%). So, compared to the sample total size, the latter is mainly composed of medium-small companies (Table 9.1).

The companies considered, the reference market, the market capitalization and the time span in which the listed companies display ESG scores are specified in Appendix (Table 9.14). According to our sample peculiarities we observe that pharmaceutical companies that globally exhibit ESG scores are only 22 on 31/12/2004, rise to 49 on 31/12/2014 and then grow considerably, just over double, to 103 on 31/12/2018. Hence, attention to ESG issues has been seeing rapid expansion in recent years, also due to the interest shown by governments of various developed and emerging economies who have incentivized and pushed companies to spread their sustainable policies. Additionally, it can be observed that almost all of the companies with the highest capitalization (+100 billion of US$) are the same that expose ESG scores since 31/12/2004.

Table 9.1 Sample distribution by market capitalization (billions of US$) on December 31, 2018

Market capitalization	*Firms number*	*Weight (%)*
0–1	20	19.4
1–5	*34*	*33.0*
5–10	19	18.4
10–30	9	8.7
30–60	5	4.9
60–100	4	3.9
100–200	7	6.8
+200	5	4.9
Total	103	100

Source Thomson Reuters Datastream; authors' elaboration

Table 9.2 Some first descriptive statistics (mean terms over the years—billions of US$)

Variable	*Obs*	*Min*	*Max*	*Mean*	*Std. Dev*
Total assets	93.67	0.02	157.10	12.60	28.97
Total debt to enterprise value	89.60	0.00	0.91	0.14	0.17
Ratio between R&D expense and net sales	82.20	0.00	359.69	5.15	40.46
Ration between capex and total assets	94.47	0.00	0.21	0.04	0.04

Source Thomson Reuters Datastream; authors' elaboration

Table 9.2 shows some first descriptive statistics relating respectively to accounting dimension,[5] financial structure,[6] innovative activity[7] and growth[8] of the 103 companies of the sample (in mean terms over the whole period of analysis). As shown in Table 9.2, the value of Total Assets, used as a proxy of firm size, has a period average of 12.6 billion of US$. In the time span considered, it is first observed that the average of total assets increases by 2.1 times and, moreover, it shows a growing trend from 2009 to 2018 (see Table 9.17 in Appendix).

The period average of the ratio between Total Debt and Enterprise Value is 0.14. The average trend in different periods is relatively unstable, recording a significant increase between 2007–2008 and 2017–2018 of 1.9 and 1.4 times respectively (see Table 9.18 in Appendix). Again, the R&D index, given by the ratio between R&D expense and Net Sales, does not have a very significant period average, due to some years showing outliers values. In particular, it is noted that the R&D index increased considerably between 2005–2006 and 2015–2016 of 18.4 and 58.4 times respectively (see Table 9.19 in Appendix). The growth indicator, approximated by the ratio between Capex and Total Assets, has a period average of 0.04. Leaving aside the 2004 average, it is observed that throughout the period considered the average is relatively stable, ranging from 0.03 to 0.04 (see Table 9.20 in Appendix).

Furthermore, Table 9.3 shows overall descriptive statistics of all the variables we are going to consider for the empirical models.

[5] Expressed in terms of Total Assets.

[6] Defined as the ratio between Total Debt and Enterprise Value.

[7] Defined as the ratio between R&D expense and Net Sales.

[8] Defined as the ratio between Capex and Total Assets.

Table 9.3 Overall descriptive statistics

Variable	*Obs*	*Min*	*Max*	*Mean*	*Std. Dev*
ESG score	720	7.78	94.18	54.05	19.46
Environment score	720	10.96	97.53	53.37	21.46
Social score	720	4.82	98.59	55.35	25.76
Governance score	720	1.17	96.22	53.28	21.71
E-resource use score	720	0.32	99.78	54.66	29.57
E-emissions score	720	0.81	99.70	54.25	29.38
E-environmental innovation score	720	0.34	99.58	51.17	26.11
S-workforce score	720	0.30	99.74	55.71	30.54
S-human rights score	720	18.49	99.06	56.68	28.15
S-community score	720	0.51	99.78	56.50	31.69
S-product responsibility score	720	2.13	99.74	52.40	29.17
G-management score	720	0.70	99.50	55.86	28.25
G-shareholders score	720	0.37	98.18	46.22	28.00
G-CSR strategy score	720	0.65	99.39	53.41	30.84
ESG controversies score	720	0.22	71.74	44.22	26.08
ESG combined score	720	7.78	87.00	43.48	13.73
ROA	1,387	−8.89	7.72	0.04	0.55
Tobin's Q	1,268	0.33	88.69	2.93	3.24
Total assets (bn of US$)	1,420	0.00047	211.62	13.19	29.85
Total debt/enterprise value	1,345	−1.01	2.11	0.14	0.18
Market to book value	1,349	−964.20	80.30	3.44	28.46
R&D/net sales	1,233	0.00	2.877.90	4.95	99.02
Capex/total assets	1,418	0.00	0.52	0.03	0.04

Source Thomson Reuters Datastream-Refinitiv; authors' elaboration

Results show that pharmaceutical companies are particularly alert to ESG factors, given that the average of the ESG Score amounts to 54.05. Instead, the ESG Controversies Score shows the lowest average at 44.22, highlighting how in the pharmaceutical sector companies are subject to scandals on average. ROA values are limited between −8.89 and 7.72, while Tobin's Q value ranging from a minimum of 0.33 to a maximum of 88.69.

The Empirical Approaches: Highlights and Results

With the purpose to verify if there is a relationship between firms ESG scores and their economic-financial performance and its impacts, we implemented empirical models on panel data.

However, in order to correctly interpret the following results, some considerations must be done first. The sample is mainly composed of medium-small companies. For the purpose of strengthening their position on the market and competing with larger companies, small- to medium-sized pharmaceutical companies are known to have the strategic objective of developing new therapies and launching a substantial number of new products on the market. Therefore, medium-small companies are used to investing significant resources in Research and Development activities.[9] On the other side, larger pharmaceutical companies are hardly competing on innovation and are increasingly interested in Merger & Acquisition operations, in order to get hold of smaller companies' inventions and consolidate their portfolios in key therapeutic areas.[10]

According to the literature and considering the sector relevance, we want to verify the following research questions: (Q_1) Is the firm performance in Pharmaceutical sector influenced by ESG scores? (Q_2) Is it possible to verify the presence of interaction between ESG scores and firm size and its impact? (Q_3) Could the relationship between ESG scores and firm performance be nonlinear?

In line with existing literature, a set of control variables was selected in order to differentiate companies in terms of size (measured by total assets), firm risk (measured by ratio between total debt and enterprise

[9] Pharmaceutical companies, in addition to invest large amounts of capital in research, in order to study and discover new drugs, have to test drugs before placing them on the market, employing significant additional investments. The patent is used to protect scientific discovery and generally has a 20 years duration, except for some drugs that can have a longer deadline. So, this long process significantly affects costs of companies that invest in research, which do not get an economic return until the drug is placed on the market.

[10] Acquiring a company that has already perfected a product is much faster than developing the same product from scratch within the company. According to different studies carried out by the worldwide largest consultancy companies, the pharmaceutical industry was constantly subjected to several M&A operations, that led over the years to the creation of the today's largest multinationals. According to Dealogic's analysis, in 2016 pharmaceutical companies, globally, spent 215 billion of US$ through the M&A tool.

value), market performance (given by the market to book value), innovative activity (measured by research and development index) and growth (using the ratio between Capex and total assets as a proxy). We focus on the overall ESG Score, the scores of its three pillars and their subcomponents, the ESG Controversies Score[11] and the ESG Combined Score.[12] As a dependent variable we consider two measure of firm performance. At first, we analyze the accounting-based measure of ROA as a proxy of firm profitability, and then Tobin's Q, a market-based measure, as a proxy of firm value in the long term. In both analyses, in order to examine the relationship between ESG scores and firm performance, a model for fixed effects panel data was used.[13]

ROA and ESG Scores

In order to analyze the relationship between ESG scores and firm profitability, the following fixed effects panel model was used:

$$\begin{aligned} \mathrm{ROA}_{it} &= \alpha_i + \beta_1 \ln(\mathrm{ESG})_{it} + \beta_2 \ln(\mathrm{Control\,Variables}) + \beta_5(\mathrm{ROA})_{it-1} \\ &\quad + D_t + u_{it}, \qquad \text{with } i = 1, \ldots, 103 \text{ and } t = 1, \ldots 15 \end{aligned}$$

where:

- The dependent variable is ROA, given by the ratio between EBITDA and book value of company's total assets. It estimates the efficiency

[11] The ESG Controversies Score measures the company exposure to environmental, social and governance controversies and negative events reflected in global media. More precisely, it reflects in a single score aspects that consider different debates that arise within each of the following area: community, human rights, management, product responsibility, use of resources, shareholders and workforce. During the year, if a scandal occurs the involved company is penalized, and this affects the overall ESGC (ESG Combined) Score and classification.

[12] The ESG Combined Score is an overall company score based on the reported information in the environment, social and governance pillars (ESG Score) with an ESG Controversies overlap acquired by global media.

[13] Following results obtained from F-test and Hausman test, in both analyses, the fixed effects model was the most appropriate to describe the observed situation, compared to Pooled OLS and random effects model respectively. In fact, some omitted variables can be linked to company's behavior belonging to the same sector, in this case the pharmaceutical one.

with which company, through its activities, generates EBITDA.[14] The choice to use a profitability index made-up by EBITDA is due to the need to make more comparable not only companies belonging to different countries, thus avoiding that estimate suffers from a different taxation, but also to be able to compare companies that adopt different investment policies.

and: the (i) Explanatory variables are, at first, the impact of ESG Score, its three pillars and their sub-components, followed by the impact of ESG Controversies Score and ESG Combined Score; (ii) Control variables are represented by total assets[15] as a proxy of company's size, by the ratio between Total Debt and Enterprise Value (EV[16]) as a leverage variable,[17] and by market to book value as a variable that measures the market performance of companies analyzed. Moreover, (iii) one-year lagged profitability was also considered among independent variables, since ROA proves to be highly correlated with its past value. In fact, firm profitability is strongly influenced not only by the sector, but also by the growth rate of new

[14] Generally, ROA is given by the ratio between EBIT and Total Assets. For this reason, the aforementioned definition of ROA can be compared to a company's EBITDA normalization on a measure of the company's accounting size. EBITDA (***Earnings Before Interest, Taxes, Depreciation and Amortization)*** represent the earning of a company before interest expense, income taxes and depreciation; it is calculated by taking the pre-tax income and adding back interest expense on debt and depreciation, depletion and amortization and subtracting interest capitalized.

[15] Following the different organizational and structural dimension of each individual company, it is appropriate to consider the company size, in order to summarize the different degree of information asymmetry and market reactivity (Lev & Thiagarajan, 1993). Most of the literature identifies total assets as a proxy of company size (Mehran, 1995).

[16] Where Enterprise Value is an accounting measure of firm total value and is defined as the sum of market capitalization, preferred stock, minority interest and total debt minus cash.

[17] Leverage represents a corporate risk that could affect future financial performance (Prior et al., 2008).

investments. (iv) We consider temporal dummies (D_t)[18]; (v) the error term u_{it} is assumed to be normally distributed, $u_{it} \sim IN(0, \sigma^2)$.

With the exception of one-year lagged ROA, in line with recent financial doctrine, all predictive variables were transformed into logarithmic form by means of natural logarithm, in order to obtain continuous values, thus avoiding distortion effects due to absolute values and reduce asymmetry in distribution.

Q_I. Firm Performance Is Influenced by ESG Scores

Table 9.4 shows regressions results. ESG Score and ESG Controversies Score are not statistically associated with firm profitability. Analyzing also the three different pillars of the overall ESG score, governance variable is the only one that is statistically associated with firm profitability, at a significance level of 5%, and shows a negative relationship with firm profitability. In quantitative terms, since the coefficient of the natural logarithm of **Governance Pillar Score** is equal to −0.021, if Governance Pillar Score increases by 10%, ROA decreases by 20 bp (*ceteris paribus*).[19]

In light of the results previously obtained, the three components of governance score were also analyzed to examine which ones have the greatest impact (Table 9.5). The result of governance variable seems to be given mainly by the management component. **Management Score** variable is the only one that is statistically significant, at a level of 1%, and shows a negative relationship with firm profitability. In particular, since the coefficient of the natural logarithm of Management Score is equal to −0.016, if Management Score increases by 10%, ROA decreases by 16 bp (*ceteris paribus*). The plausible interpretation of a negative relationship is that costs associated with governance commitment can have a negative

[18] Temporal dummies are binary variables relating to individual years of the observation period, for a total of 15 temporal dummies (from 2004 to 2018). These variables are very important, since they allow to emphasize the existing relationship among different years considered and capture the common trend effect to all companies on their economic-financial performance. In fact, not only internal factors but also external factors impact on corporate performance.

[19] Dependent variable increases more slowly than independent variable, since their relationship is expressed in half-logarithmic form. In formula: $\Delta\text{ROA} = \beta_1 + \ln\left(1 + \frac{\Delta\text{ESG}}{\text{ESG}}\right) \rightarrow$ $0.021 * \ln(1+0.1) = 0.0020$. The change of dependent variable is transformed in basis points (1 bp = 0.01%).

Table 9.4 ROA and ESG scores

	4.1	*4.2*	*4.3*	*4.4*	*4.5*	*4.6*
ESG score	0.0004 (0.0201)					
ESG controversies score					0.0041 (0.0051)	
ESG combined score						0.0074 (0.0159)
Environment score		0.0244 (0.0158)				
Social score			0.0185 (0.0129)			
Governance score				**−0.0212**** (0.0102)		
Total assets	−0.0122 (0.0101)	−0.0144 (0.0101)	−0.0153 (0.0102)	−0.0118 (0.0099)	−0.0107 (0.0101)	−0.0122 (0.0100)
D/EV	−0.0090** (0.0039)	−0.0089** (0.0039)	−0.0094** (0.0039)	−0.0084** (0.0039)	−0.0090** (0.0039)	−0.0091** (0.0039)
Market to book value	−0.0043 (0.0094)	−0.0028 (0.0094)	−0.0045 (0.0094)	−0.0036 (0.0094)	−0.0045 (0.0094)	−0.0042 (0.0094)
ROA (t − 1)	0.3061*** (0.0405)	0.3059*** (0.0403)	0.3020*** (0.0405)	0.3037*** (0.0403)	0.3034*** (0.0405)	0.3046*** (0.0406)
Constant	0.3465 (0.2391)	0.2969 (0.2358)	0.3419 (0.2337)	0.4244* (0.2361)	0.3030 (0.2402)	0.3204 (0.2412)
Temporal dummies	Yes	Yes	Yes	Yes	Yes	Yes
R-squared LSDV	0.7829	0.7839	0.7837	0.7847	0.7831	0.7830
Observations	634	634	634	634	634	634
Firms number	103	103	103	103	103	103

***, **, and * indicate a statistical significance level of 1%, 5% and 10% respectively. Standard errors are shown in brackets. All explanatory variables are expressed in logarithmic form through the natural logarithm, except for one-year lagged profitability. Dependent variable ROA $= \frac{\text{EBITDA}}{\text{Total Asset}}$. Fixed effects model

Source Thomson Reuters Datastream-Refinitiv; authors' elaboration

impact on profitability, in this case, especially costs to be incurred in order to pursue best practices of corporate governance principles.

Overall ESG measure, as well as its three pillars, are generally less associated with firm performance than their individual components. Therefore, especially in light of the sample heterogeneity, the impact

Table 9.5 ROA and components score of Governance Pillar Score variable

	5.1	*5.2*	*5.3*
Management score	−0.0164*** (0.0061)		
Shareholders score		−0.0026 (0.0055)	
CSR strategy score			0.0015 (0.0084)
Total assets	−0.0116 (0.0100)	−0.0124 (0.0100)	−0.0126 (0.0102)
D/EV	−0.0089** (0.0039)	−0.0089** (0.0039)	−0.0090** (0.0039)
Market to book value	−0.0036 (0.0093)	−0.0044 (0.0094)	−0.0043 (0.0094)
ROA ($t-1$)	0.3107*** (0.0402)	0.3039*** (0.0407)	0.3064*** (0.0405)
Constant	0.3987* (0.2333)	0.3616 (0.2360)	0.3502 (0.2347)
Temporal dummies	Yes	Yes	Yes
R-squared LSDV	0.7858	0.7830	0.7829
Observations	634	634	634
Firms number	103	103	103

***, **, and * indicate a statistical significance level of 1%, 5% and 10% respectively. Standard errors are shown in brackets. All explanatory variables are expressed in logarithmic form through the natural logarithm, except for one-year lagged profitability. Dependent variable ROA $= \frac{\text{EBITDA}}{\text{Total Asset}}$. Fixed effects model
Source Thomson Reuters Datastream-Refinitiv; authors' elaboration

on firm profitability of the individual sub-components that make up the environmental and social pillar was also analyzed (Tables 9.6 and 9.7).

Among the three different sub-components that make up the environmental pillar, **Emissions Score** is the only variable that is statistically significant, at a level of 1%, and shows a positive relationship with firm profitability. A greater pharmaceutical companies' commitment in order to reduce polluting emissions into the atmosphere, seems to have a positive impact on firm profitability. In quantitative terms, since the coefficient of the natural logarithm of Emissions Score is equal to 0.028, if Emissions Score increases by 10%, ROA increases by 27 bp (*ceteris paribus*).

Instead, among the four different sub-components that make up the social pillar, none is statistically significant in the profitability analysis.

Table 9.6 ROA and components score of Environment Pillar Score variable

	6.1	*6.2*	*6.3*
Resource use score	−0.0007		
	(0.0071)		
Emissions score		**0.0281***	
		(0.0065)	
Environmental innovation score			−0.0008
			(0.0064)
Total assets	−0.0120	−0.0157	−0.0123
	(0.0101)	(0.0098)	(0.0100)
D/EV	−0.0090**	−0.0101***	−0.0090**
	(0.0039)	(0.0039)	(0.0039)
Market to book value	−0.0044	−0.0002	−0.0042
	(0.0095)	(0.0093)	(0.0094)
ROA (t − 1)	0.3060***	0.2931***	0.3061***
	(0.0404)	(0.0398)	(0.0404)
Constant	0.3461	0.3029	0.3531
	(0.2344)	(0.2302)	(0.2383)
Temporal dummies	Yes	Yes	Yes
R-squared LSDV	0.7829	0.7904	0.7829
Observations	634	634	634
Firms number	103	103	103

***, **, and * indicate a statistical significance level of 1%, 5% and 10% respectively. Standard errors are shown in brackets. All explanatory variables are expressed in logarithmic form through the natural logarithm, except for one-year lagged profitability. Dependent variable ROA $= \frac{\text{EBITDA}}{\text{Total Asset}}$. Fixed effects model

Source Thomson Reuters Datastream-Refinitiv; authors' elaboration

Q_2. There Is Interaction Between ESG Scores and Firm Size

Given the shortage of overall results previously obtained, it was considered appropriate to estimate the interaction of ESG variables with firm size (Table 9.8). It is reasonable to think that, by adding an interaction term between firm size (approximated by total assets)[20] and different ESG variables, further results can be achieved.

First of all, **Environment Pillar Score** variable is now statistically significant at a level of 5% and shows a positive relationship with firm

[20] Total assets, used as a proxy of firm size, are not necessarily a proxy of the firm's capitalization change.

Table 9.7 ROA and components score of Social Pillar Score variable

	7.1	*7.2*	*7.3*	*7.4*
Workforce score	0.0041			
	(0.0057)			
Human rights score		0.0053		
		(0.0132)		
Community score			0.0007	
			(0.0079)	
Product responsibility score				0.0090
				(0.0077)
Total assets	−0.0134	−0.0127	−0.0122	−0.0138
	(0.0101)	(0.0100)	(0.0100)	(0.0101)
D/EV	−0.0093**	−0.0089**	−0.0090**	−0.0090**
	(0.0039)	(0.0039)	(0.0039)	(0.0039)
Market to book value	−0.0047	−0.0039	−0.0043	−0.0047
	(0.0094)	(0.0094)	(0.0094)	(0.0094)
ROA ($t - 1$)	0.3051***	0.3055***	0.3060***	0.3026***
	(0.0404)	(0.0404)	(0.0405)	(0.0405)
Constant	0.3591	0.3372	0.3464	0.3492
	(0.2345)	(0.2354)	(0.2343)	(0.2338)
Temporal dummies	Yes	Yes	Yes	Yes
R-squared LSDV	0.7831	0.7829	0.7829	0.7834
Observations	634	634	634	634
Firms number	103	103	103	103

***, **, and * indicate a statistical significance level of 1%, 5% and 10% respectively. Standard errors are shown in brackets. All explanatory variables are expressed in logarithmic form through the natural logarithm, except for one-year lagged profitability. Dependent variable ROA $= \frac{\text{EBITDA}}{\text{Total Asset}}$. Fixed effects model

Source Thomson Reuters Datastream-Refinitiv; authors' elaboration

profitability. In particular, since the coefficient of the natural logarithm of Environment Pillar Score is equal to 0.375, if Environment Pillar Score increases by 10%, ROA increases by 357 bp (*ceteris paribus*). However, the interaction term between firm size and environmental score is negative, thus suggesting that as firm size increases, there is a negative relationship between profitability and Environmental Pillar Score variable. Evidently, larger companies have to face higher costs in order to engage environmental issues, such as reduction of resources use and emissions, and these higher costs therefore seem to penalize profitability.

Governance Pillar Score variable maintains the same statistical significance, at a level of 5%, and also the same negative relationship with firm profitability. In quantitative terms, if Governance Pillar Score increases

Table 9.8 ROA and interaction between ESG scores and firm size

	8.1	*8.2*	*8.3*	*8.4*	*8.5*	*8.6*
ESG score	0.0232 (0.1980)					
ESG controversies score					**−0.2242***** (0.0791)	
ESG combined score						−0.1927 (0.1710)
Environment score		**0.3749**** (0.1689)				
Social score			0.0956 (0.1531)			
Governance score				**−0.2392**** (0.1184)		
Total assets	−0.0080 (0.0371)	0.0481 (0.0316)	−0.0007 (0.0306)	−0.0460** (0.0210)	−0.0439*** (0.0152)	−0.0440 (0.0288)
Total assets * ESG score	−0.0011 (0.0093)					
Total assets * ESG controversies score					**0.0094***** (0.0032)	
Total assets * ESG combined score						0.0091 (0.0077)
Total assets * environment score		**−0.0160**** (0.0077)				
Total assets * social score			−0.0037 (0.0072)			
Total assets * governance score				**0.0102*** (0.0055)		
D/EV	−0.0090** (0.0039)	−0.0085** (0.0039)	−0.0095** (0.0039)	−0.0091** (0.0039)	−0.0086** (0.0039)	−0.0092** (0.0039)
Market to book value	−0.0045 (0.0096)	−0.0050 (0.0095)	−0.0050 (0.0094)	−0.0003 (0.0095)	−0.0065 (0.0094)	−0.0032 (0.0094)
ROA ($t-1$)	0.3057*** (0.0407)	0.3056*** (0.0402)	0.2998*** (0.0407)	0.3061*** (0.0402)	0.3127*** (0.0404)	0.3081*** (0.0407)
Constant	0.2600 (0.7822)	−1.0561 (0.6904)	0.0383 (0.6448)	1.1309** (0.4489)	1.1218*** (0.3700)	1.0148 (0.6383)
Temporal dummies	Yes	Yes	Yes	Yes	Yes	Yes
R-squared LSDV	0.7829	0.7857	0.7838	0.7861	0.7866	0.7835
Observations	634	634	634	634	634	634
Firms number	103	103	103	103	103	103

***, **, and * indicate a statistical significance level of 1%, 5% and 10% respectively. Standard errors are shown in brackets. Firm size is approximated by the book value of total assets. All explanatory variables are expressed in logarithmic form through the natural logarithm, except for one-year lagged profitability. Dependent variable ROA $= \frac{\text{EBITDA}}{\text{Total Asset}}$. Fixed effects model

Source Thomson Reuters Datastream-Refinitiv; authors' elaboration

by 10%, ROA decreases by 228 bp (*ceteris paribus*). However, from the interaction with firm size, it emerges that as firm size increases, there is a positive relationship between profitability and governance score. Adopting corporate governance best practices would therefore seem to be more expensive for smaller companies.

Lastly, **ESG Controversies Score** variable also acquires statistical significance, at a level of 1%, and shows a negative relationship with firm profitability. In quantitative terms, if ESG Controversies Score increases by 10%, ROA decreases by 214 bp (*ceteris paribus*). However, the interaction term between firm size and controversies score is positive, thus suggesting that as firm size increases, there is a positive relationship between profitability and controversies score. The more virtuous a company is considered, the more likely appreciated by stakeholders, especially when these virtues are disclosed by external sources such as global media. Since larger companies (those best known to stakeholders) are more exposed to corporate scandals on average, their virtuous behavior is likely to attract a greater media and stakeholder's attention, thus positively impacting on profitability. Smaller pharmaceutical companies' profitability seems, instead, to be penalized by an increase in controversies score, presumably due to higher costs that have to be incurred in order to guarantee socially responsible behavior and avoid any potential scandal, thus keeping intact their own reputation (from product responsibility, management, shareholders, human rights, workforce, resource use, responsibility toward community).

Finally, a robustness test was performed in search of a nonlinear relationship between ESG scores and firm profitability (Q3). With this purpose the quadratic term of ESG scores was added to regressions. However, despite results suggest the existence of a nonlinear relationship, none of ESG variables is statistically significant in the profitability analysis.

Tobin's Q and ESG Scores

In order to analyze the relationship between ESG scores and firm value in terms of Tobin's Q, the following fixed effects panel model was used:

$$\ln(\text{Tobin}'\text{s Q})_{it} = \alpha_i + \beta_1 \ln(\text{ESG})_{it-1} + \beta_2 \ln(\text{Control Variables}) + D_t + u_{it}, \quad \text{with}\, i = 1, \ldots, 103\, \text{and}\, t = 1, \ldots 15$$

where:

- The dependent variable is represented by the natural logarithm of Tobin's Q, as a proxy of firm value or its profitability prospects, as it is intended to capture the value of the long-term investments including intangibles.[21] It preferred to use the logarithmic form because through this transformation it is possible to harmonize the linear effect of the relationship.[22] Considering various approximations available in literature of Tobin's Q, the one originally provided by Kaplan and Zingales (1997) was used. In formula: Tobin′s Q = $\frac{TA+MVE-(BVE+DT)}{TA}$,[23] where TA is the book value of company's total assets, MVE is the market value of equity,[24] BVE is the book value of equity,[25] and DT are deferred taxes in financial statement.[26]

and: the (i) Explanatory variables are, at first, the impact of ESG Score, its three pillars and their sub-components, followed by the impact of ESG Controversies Score and ESG Combined Score; (ii) Control variables are represented by the ratio between total debt and enterprise value as a leverage variable, by the research and development index (i.e. the ratio

[21] Tobin's Q adequately captures the effect of specific intangible assets (Morck et al., 1988) and typical financial and market variables (Brown & Caylor, 2004).

[22] Furthermore, the choice to use a logarithmic form for the dependent variable, proves to be more consistent with the assumption that error terms are normally distributed in regressions (Greene, 2000).

[23] Kaplan and Zingales (1997) measure Tobin's Q as the ratio between market value of assets and book value of total assets; where the market value of assets is approximated by the sum of book value of assets and market value of equity minus the sum of book value of equity and deferred taxes. They argue that their formulation (particularly when Tobin's Q is the dependent variable in fixed effects regressions) approximate Tobin's Q better than other mathematically more complex and representative measures, like the one provided by Chung and Pruitt (1994) and Perfect and Wiles (1994).

[24] I.e. the market value of shares multiplied by the number of ordinary shares outstanding.

[25] I.e. the common shareholders' investment in a company.

[26] I.e. the accumulation of taxes which are deferred as a result of timing differences between sales and expenses for tax and financial reporting purposes.

between R&D expense[27] and net sales or revenues[28]) as an innovative activity variable, and the ratio between Capex[29] and total assets[30] as a growth variable for companies analyzed. Furthermore: (iii) We consider temporal dummies (D_t)[31]; (iv) the error term u_{it} is assumed to be normally distributed, $u_{it} \sim IN(0, \sigma^2)$.

In line with recent financial doctrine, all predictive variables were transformed into logarithmic form by means of natural logarithm, in order to obtain continuous values and reduce asymmetry in distribution. In addition, in line with existing literature, these variables were all lagged by one year, since they proved to have a greater impact on firm value in the following period.

Q_I. Firm Performance Is Influenced by ESG Scores

Regressions results are shown in Table 9.9. ESG Score and ESG Controversies Score are not statistically associated with firm value.

Analyzing also the three different pillars of the overall ESG score, governance variable is the only one that is statistically associated with firm value. In particular, it is statistically significant at a level of 5% and shows a positive relationship with firm value. In quantitative terms, since the coefficient of the natural logarithm of **Governance Pillar Score** is equal to 0.070, if Governance Pillar Score at time $t-1$ increases by 10%, Tobin's Q at time t increases by 0.70% (*ceteris paribus*).[32]

[27] It represents all direct and indirect costs related to the creation and development of new processes, techniques, applications and products with commercial possibilities.

[28] R&D expense not only affect long-term financial performance but also social performance, given that many elements of corporate social responsibility are linked to product and/or process innovation (McWilliams & Siegel, 2000).

[29] Capital Expenditures.

[30] I.e. how a company uses its activities to invest in operating fixed assets.

[31] Temporal dummies are binary variables relating to individual years of the observation period, for a total of 15 temporal dummies (from 2004 to 2018). These variables are very important, since they allow to emphasize the existing relationship among different years considered and capture the common trend effect to all companies on their economic-financial performance. In fact, not only internal factors but also external factors impact on corporate performance.

[32] The relationship between dependent and independent variable is linear, since both variables are expressed in logarithmic form. In formula: $\frac{\Delta Q}{Q} = \beta_1 + \frac{\Delta \text{ESG}}{\text{ESG}} \rightarrow 0.070 * 0.1 = 0.0070$.

Table 9.9 Tobin's Q and ESG scores

	9.1	*9.2*	*9.3*	*9.4*	*9.5*	*9.6*
ESG score ($t-1$)	0.0288 (0.0658)					
ESG controversies score ($t-1$)					0.0001 (0.0148)	
ESG combined score ($t-1$)						0.0194 (0.0521)
Environment score ($t-1$)		**−0.0868*** (0.0508)				
Social score ($t-1$)			0.0352 (0.0440)			
Governance score ($t-1$)				**0.0701**** (0.0318)		
D/EV ($t-1$)	−0.0538*** (0.0132)	−0.0507*** (0.0131)	−0.0541*** (0.0131)	−0.0578*** (0.0132)	−0.0530*** (0.0131)	−0.0535*** (0.0131)
R&D/sales ($t-1$)	−0.0400 (0.0342)	−0.0340 (0.0339)	−0.0410 (0.0341)	−0.0471 (0.0340)	−0.0379 (0.0339)	−0.0394 (0.0342)
Capex/assets ($t-1$)	0.0455* (0.0239)	0.0480** (0.0238)	0.0450* (0.0239)	0.0458* (0.0237)	0.0464* (0.0240)	0.0452* (0.0240)
Constant	0.6152* (0.3239)	1.1227*** (0.2596)	0.5814** (0.2444)	0.4349** (0.1943)	0.7437*** (0.1454)	0.6609** 0.2614
Temporal dummies	Yes	Yes	Yes	Yes	Yes	Yes
R-squared LSDV	0.8381	0.8391	0.8382	0.8398	0.8380	0.8381
Observations	538	538	538	538	538	538
Firms number	103	103	103	103	103	103

***, **, and * indicate a statistical significance level of 1%, 5% and 10% respectively. Standard errors are shown in brackets. All explanatory variables are expressed in logarithmic form through the natural logarithm. Dependent variable Tobin's $Q = \frac{TA+MVE-(BVE+DT)}{TA}$. Fixed effects model
Source Thomson Reuters Datastream-Refinitiv; authors' elaboration

Environment Pillar Score variable is statistically significant at a level of 10% and shows a negative relationship with firm value. However, since its *p*-value is greater than 0.05, at a significance level of 5%, the null

Table 9.10 Tobin's Q and components score of Governance Pillar Score variable

	10.1	*10.2*	*10.3*
Management score ($t-1$)	**0.0464**** (0.0198)		
Shareholders score ($t-1$)		−0.0119 (0.0168)	
CSR strategy score ($t-1$)			−0.0179 (0.0267)
D/EV ($t-1$)	−0.0555*** (0.0130)	−0.0519*** (0.0132)	−0.0520*** (0.0132)
R&D/sales ($t-1$)	−0.0447 (0.0338)	−0.0375 (0.0339)	−0.0348 (0.0342)
Capex/assets ($t-1$)	0.0458* (0.0237)	0.0470** (0.0238)	0.0441* (0.0241)
Constant	0.5404*** (0.1603)	0.7917*** (0.1512)	0.8135*** (0.1704)
Temporal dummies	Yes	Yes	Yes
R-squared LSDV	0.8400	0.8382	0.8382
Observations	538	538	538
Firms number	103	103	103

***, **, and * indicate a statistical significance level of 1%, 5% and 10% respectively. Standard errors are shown in brackets. All explanatory variables are expressed in logarithmic form through the natural logarithm. Dependent variable Tobin's Q = $\frac{TA+MVE-(BVE+DT)}{TA}$. Fixed effects model

Source Thomson Reuters Datastream-Refinitiv; authors' elaboration

hypothesis, according to which the coefficient of Environment Pillar Score variable is equal to 0, cannot be rejected.[33]

In light of the results previously obtained, the three components of governance score were also analyzed to examine which ones have the greatest impact (Table 9.10).

The result of governance variable seems to be given mainly by the management component. **Management Score** variable is the only one that is statistically significant, at a level of 5%, and shows a positive relationship with firm value. In particular, since the coefficient of the natural logarithm of Management Score is equal to 0.046, if Management Score

[33] A large p-value indicates that there is weak evidence in data for rejecting the null hypothesis (i.e., H_0: β=0). By convention, the threshold value is usually 0.05.

at time $t - 1$ increases by 10%, Tobin's Q at time t increases by 0.46% (*ceteris paribus*). The plausible interpretation of a positive relationship is that governance commitment, especially effort to ensure that companies effectively pursue best practices of corporate governance principles, appears to reduce replacement costs of the capital stock,[34] thus increasing Tobin's Q value. One reason for this positive impact can be related to the fact that skills and professional experience of directors contribute to a better implementation of corporate strategies through advanced decision-making processes, in terms of efficiency and effectiveness.

Overall ESG measure, as well as its three pillars, are generally less associated with firm performance than their individual components. Therefore, especially in light of the sample heterogeneity, the impact on firm value of the individual sub-components that make up the environmental and social pillar was also analyzed (Tables 9.11 and 9.12).

Among the three different sub-components that make up the environmental pillar, **Environmental Innovation Score** variable is statistically significant at a level of 5% and shows a positive relationship with firm value.

In quantitative terms, since the coefficient of the natural logarithm of Environmental Innovation Score is equal to 0.039, if Environmental Innovation Score at time $t - 1$ increases by 10%, Tobin's Q at time t increases by 0.39% (*ceteris paribus*). Instead, **Resource Use Score** variable is statistically significant at a level of 10% and shows a negative relationship with firm value. However, since its p-value is greater than 0.05, at a significance level of 5%, the null hypothesis, according to which the coefficient of Resource Use Score variable is equal to 0, cannot be rejected.

Among the four different sub-components that make up the social pillar, **Community Score** is the only variable that is statistically significant, at a level of 1%, and shows a positive relationship with firm value. The pharmaceutical companies' contribution to the community, toward being a good citizen, protecting public health and respecting business ethics, seems to be rewarded by the market, thus positively impacting on firm value. In quantitative terms, since the coefficient of the natural logarithm of Community Score is equal to 0.070, if Community Score at time $t - 1$ increases by 10%, Tobin's Q at time t increases by 0.70% (*ceteris paribus*).

[34] Replacement costs of capital stock (assets replacement costs) are, in this case, approximated by the book value of company's total assets.

Table 9.11 Tobin's Q and components score of Environment Pillar Score variable

	11.1	*11.2*	*11.3*
Resource use score ($t-1$)	**−0.0391***		
	(0.0207)		
Emissions score ($t-1$)		−0.0272	
		(0.0202)	
Environmental innovation score ($t-1$)			**0.0393****
			(0.0186)
D/EV ($t-1$)	−0.0505***	−0.0516***	−0.0541***
	(0.0131)	(0.0131)	(0.0130)
R&D/sales ($t-1$)	−0.0294	−0.0453	−0.0437
	(0.0341)	(0.0343)	(0.0339)
Capex/assets ($t-1$)	0.0462*	0.0450*	0.0443*
	(0.0237)	(0.0238)	(0.0237)
Constant	0.9325***	0.8387***	0.5748***
	(0.1676)	(0.1523)	(0.1567)
Temporal dummies	Yes	Yes	Yes
R-squared LSDV	0.8393	0.8387	0.8396
Observations	538	538	538
Firms number	103	103	103

***, **, and * indicate a statistical significance level of 1%, 5% and 10% respectively. Standard errors are shown in brackets. All explanatory variables are expressed in logarithmic form through the natural logarithm. Dependent variable Tobin's $Q = \frac{TA+MVE-(BVE+DT)}{TA}$. Fixed effects model
Source Thomson Reuters Datastream-Refinitiv; authors' elaboration

Q_2. There Is Interaction Between ESG Scores and Firm Size

Given the shortage of overall results previously obtained, it was considered appropriate to estimate the interaction of ESG variables with firm size (Table 9.13). It is reasonable to think that, by adding an interaction term between firm size (approximated by book value of total assets) and different ESG variables, further results can be achieved.

First of all, both **ESG Score** and **Social Pillar Score** are now statistically significant, at a level of 1% and 5% respectively, and show a positive relationship with firm value. In quantitative terms, since the coefficient of the natural logarithm of ESG Score is equal to 2.621, if ESG Score at time $t-1$ increases by 10%, Tobin's Q at time t increases by 26.2% (*ceteris paribus*). This result seems to be driven mainly by governance variable, also statistically significant at 1% and positively correlated to firm value. In

Table 9.12 Tobin's Q and components score of Social Pillar Score variable

	12.1	*12.2*	*12.3*	*12.4*
Workforce score ($t-1$)	0.0254			
	(0.0184)			
Human rights score ($t-1$)		−0.0574		
		(0.0411)		
Community score ($t-1$)			**0.0703*** **	
			(0.0248)	
Product responsibility score ($t-1$)				0.0274
				(0.0232)
D/EV ($t-1$)	−0.0553***	−0.0536***	−0.0550***	−0.0528***
	(0.0131)	(0.0130)	(0.0130)	(0.0130)
R&D/sales ($t-1$)	−0.0432	−0.0356	−0.0395	−0.0364
	(0.0341)	(0.0339)	(0.0336)	(0.0339)
Capex/assets ($t-1$)	0.0446*	0.0454*	0.0500**	0.0466*
	(0.0238)	(0.0238)	(0.0236)	(0.0238)
Constant	0.6132***	0.9863***	0.4627***	0.6383***
	(0.1653)	(0.2196)	(0.1669)	(0.1621)
Temporal dummies	Yes	Yes	Yes	Yes
R-squared LSDV	0.8387	0.8387	0.8409	0.8385
Observations	538	538	538	538
Firms number	103	103	103	103

***, **, and * indicate a statistical significance level of 1%, 5% and 10% respectively. Standard errors are shown in brackets. All explanatory variables are expressed in logarithmic form through the natural logarithm. Dependent variable Tobin′s $Q = \frac{TA+MVE-(BVE+DT)}{TA}$. Fixed effects model
Source Thomson Reuters Datastream-Refinitiv; authors' elaboration

fact, if **Governance Pillar Score** at time $t-1$ increases by 10%, Tobin's Q at time t increases by 22.3% (*ceteris paribus*). While a 10% increase in Social Pillar Score at time $t-1$ leads to a 12.4% increase in Tobin's Q at time t (*ceteris paribus*). However, interaction terms between firm size and aforementioned ESG scores are negative, thus suggesting that as the firm size increases, there is a negative relationship with Tobin's Q. So, in firm value terms, it would seem convenient for smaller companies to engage ESG issues (in particular social and governance issues), as this greater commitment would seem to reduce replacement costs of capital stock.

ESG Controversies Score variable also acquires statistical significance, at a level of 1%, and shows a negative relationship with firm value. In quantitative terms, if ESG Controversies Score at time $t-1$ increases by 10%, Tobin's Q at time t decreases by 13.3% (*ceteris paribus*). However,

Table 9.13 Tobin's Q and interaction between ESG scores and firm size

	13.1	*13.2*	*13.3*	*13.4*	*13.5*	*13.6*
ESG score (t – 1)	**2.6211*** **					
	(0.6436)					
ESG controversies score (t – 1)					**–1.3317*** **	
					(0.2673)	
ESG combined score (t – 1)						**0.9491* **
						(0.5552)
Environment score (t – 1)		0.7398				
		(0.5712)				
Social score (t – 1)			**1.2434** **			
			(0.4851)			
Governance score (t – 1)				**2.2323*** **		
				(0.3625)		
Total assets (t – 1)	0.3025***	0.0022	0.0601	0.1946***	–0.3497***	0.0054
	(0.1162)	(0.1019)	(0.0935)	(0.0635)	(0.0528)	(0.0912)
Total assets * ESG score (t – 1)	**–0.1192*** **					
	(0.0299)					
Total assets * ESG controversies score (t – 1)					**0.0540*** **	
					(0.0109)	
Total assets * ESG combined score (t – 1)						**–0.0418* **
						(0.0248)
Total assets * environment score (t – 1)		–0.0360				
		(0.0256)				
Total assets * social score (t – 1)			**–0.0547** **			
			(0.0227)			

(continued)

Table 9.13 (continued)

	13.1	*13.2*	*13.3*	*13.4*	*13.5*	*13.6*
Total assets * governance score ($t-1$)				**−0.1010*****		
				(0.0168)		
D/EV ($t-1$)	−0.0379***	−0.0390***	−0.0425***	−0.0418***	−0.0333**	−0.0424***
	(0.0130)	(0.0131)	(0.0130)	(0.0127)	(0.0129)	(0.0131)
R&D/sales ($t-1$)	−0.0358	−0.0354	−0.0439	−0.0567*	−0.0257	−0.0446
	(0.0332)	(0.0334)	(0.0333)	(0.0321)	(0.0326)	(0.0335)
Capex/assets ($t-1$)	−0.0267	−0.0028	−0.0176	−0.0105	0.0069	−0.0082
	(0.0261)	(0.0264)	(0.0262)	(0.0244)	(0.0247)	(0.0264)
Constant	−5.8703**	0.9503	−0.7554	−3.5080**	9.1175***	0.5027
	(2.4934)	(2.2694)	(1.9994)	(1.3706)	(1.2777)	(2.0696)
Temporal dummies	Yes	Yes	Yes	Yes	Yes	Yes
R-squared LSDV	0.8502	0.8456	0.8475	0.8577	0.8529	0.8454
Observations	538	538	538	538	538	538
Firms number	103	103	103	103	103	103

***, **, and * indicate a statistical significance level of 1%, 5% and 10% respectively. Standard errors are shown in brackets. All explanatory variables are expressed in logarithmic form through the natural logarithm. Firm size is approximated by the book value of total assets. Dependent variable Tobin′s $Q = \frac{TA+MVE-(BVE+DT)}{TA}$. Fixed effects model

Source Thomson Reuters Datastream-Refinitiv; authors' elaboration

the interaction term between firm size and controversies score is positive, thus suggesting that as firm size increases, there is a positive relationship between firm value and controversies score. The more virtuous a company is considered, the more likely appreciated by stakeholders, especially when these virtues are disclosed by external sources such as global media. Since larger companies (those best known to stakeholders) are more exposed to corporate scandals on average, their virtuous behavior is likely to attract a greater media and stakeholder's attention, thus positively impacting on firm value. Smaller pharmaceutical companies' value seems, instead, to be penalized by an increase in controversies score, presumably increasing replacement costs of companies' activities.

Lastly, **ESG Combined Score** variable is now statistically significant at a level of 10% and shows a positive relationship with firm value. However, since its *p*-value is greater than 0.05, at a significance level of 5%, the null hypothesis, according to which the coefficient of ESG Combined Score is equal to 0, cannot be rejected.

Finally, a robustness test was performed in search of a nonlinear relationship between ESG scores and firm value (Q_3). We added the quadratic term of ESG scores. The output results do not change underling the idea that the governance score continues to be the most relevant in determining firm value.

ESG Score, **Social Pillar Score**, **Governance Pillar Score** and **ESG Combined Score** variables are statistically significant (at 1%, 5%, 1% and 5% respectively) and show an inverted U-shaped nonlinear relationship with Tobin's Q, since their quadratic term coefficient is negative. This inverse relationship suggests that when these ESG scores are small, firm value increases but, at some point, as ESG scores increases, firm value decreases. The plausible interpretation of this result is that a lower commitment to ESG issues, especially social and governance ones, would seem to be initially convenient for companies (probably causing a reduction in replacement costs of capital stock), but, at some point, if the commitment to social and governance issues increases, this translates into a firm value reduction (presumably increasing replacement costs of assets).

Therefore, there would be a threshold value of these ESG scores beyond which it would be inconvenient for companies to invest additional resources in sustainable activities. However, this result could be due to the limited nature of this time series (15 years), given some peculiarities of the pharmaceutical sector. In fact, considering the huge research and development investments of companies making up the sample, and having their patents generally a 20 years duration, it is possible to hypothesize that for these companies investing additional high resources also in sustainable activities requires a greater availability of time series data than the one currently considered; so that a greater investment can be adequately captured and positively impact on firm value.

Conclusions

This chapter analyzed the impact of ESG scores on firm performance, using a sample of 103 worldwide listed pharmaceutical companies in the period from 2004 to 2018.[35] Explanatory variables on which research questions have been addressed are: ESG Score, its three pillars and their sub-components, ESG Controversies Score and ESG Combined Score. Results obtained are different depending on the firm performance measure used. As a performance measure was considered at first the ROA index, an accounting-based measure, as a proxy of firm profitability, and then Tobin's Q, a market-based measure, as a proxy of firm value in long term. The most statistically stable results were obtained when Tobin's Q was used, as it is a measure intended to capture long-term investments value including intangibles, considering both *p*-value and good model adaption (R-squared LSDV). In summary, we tested the following research Hypothesis: (H_1) firm performance is influenced by ESG score; (H_2) there is interaction between ESG scores and firm size; (H_3) the relationship between ESG scores and firm performance is nonlinear.

To whom it concerns the first research question, the main result of the empirical analysis is the existence of a statistically significant relationship between Governance Pillar Score and firm performance. In particular, this result seems to be driven mainly by Management Score component, which is also negatively associated with ROA and positively with Tobin's Q. So, adopting best practices to corporate governance principles seems to damage firm profitability, presumably due to high costs that have to be incurred but, in long term, it appears to be an advantageous choice, capable of increasing firm value. One reason for this positive impact can be motivated by the fact that directors' skills and professional experience contribute to a better implementation of corporate strategies through advanced decision-making processes, in terms of

[35] The sample is mainly composed of small to medium sized companies.

efficiency and effectiveness. In quantitative terms, it was found that a 10% increase in Governance Pillar Score leads to a decrease in ROA of 20 bp and an increase in Tobin's Q at time $t + 1$ of 0.70% (*ceteris paribus*). Instead, a 10% increase in Management Score implies a 16 bp reduction in profitability and a 0.46% increase in firm value at time $t + 1$ (*ceteris paribus*).

Furthermore, firm size appears extremely significant since, adding an interaction term between firm size (approximated by total assets) and ESG scores, all variables on which research hypotheses were tested, acquired statistical significance. In particular, it emerged that: (a) interaction between overall ESG score and firm size is negatively associated with Tobin's Q; (b) interaction between environmental score and firm size is negatively associated with ROA; (c) interaction between social score and firm size is negatively associated with Tobin's Q; (d) interaction between governance score and firm size is positively associated with ROA and negatively with Tobin's Q; (e) interaction between ESG controversies score and firm size is positively associated with both ROA and Tobin's Q; (f) interaction between ESG combined score and firm size is negatively associated with Tobin's Q. Therefore, the output results highlighted that for smaller companies there is a negative relationship between firm profitability with both governance and ESG controversies score. Since small-to medium-sized pharmaceutical companies incur massive costs in research and development activities, it is reasonable to deduce that for these companies, investing additional resources in these ESG factors can damage their profitability. However, from a long-term perspective, for smaller companies there is a positive relationship between the impact of overall ESG score, mainly driven by social and governance one, and firm value. The main hypothesis that supports the positive relationship is that attention to sustainable issues improves companies' competitiveness and reputation, and the market would seem to reward small-to medium-sized companies, those that more invest in innovation. In quantitative terms, if ESG Score increases by 10%, Tobin's Q at time $t + 1$ increases by 26.2% (*ceteris paribus*). Instead, a 10% increase in Governance Pillar Score implies a 22.3% increase in firm value at time $t + 1$ (*ceteris paribus*). Lastly, it emerged that as firm size increases, there is a positive relationship between ESG controversies not only with ROA, but also with Tobin's Q. Since larger companies (those best known to stakeholders) are more exposed to corporate scandals on average, their virtuous behavior is likely to attract a greater media and stakeholder's attention, thus positively impacting on firm performance. On the contrary, for smaller companies, a greater media and stakeholder's attention to corporate controversies is likely to trigger a greater skepticism toward company, thus leading to a lower credibility and a consequent firm performance reduction. In quantitative terms, a 10% increase in ESG Controversies Score leads to a ROA reduction of 214 bp and a Tobin's Q reduction at time $t + 1$ of 13.3% (*ceteris paribus*).

At last, we point out an inverted U-shaped nonlinear relationship between ESG Score, Social Pillar Score, Governance Pillar Score and ESG Combined Score with Tobin's Q.[36] This inverse relationship suggests that when these ESG scores are small, firm value increases but, at some point, as these scores increase, firm value decreases. So, a companies' increased investment in sustainable activities, especially in social and governance ones, seems to penalize them with a lower firm value in long term.

However, this last result could be due to the limited nature of this time series (15 years),[37] given some peculiarities of the pharmaceutical sector. In fact, considering the huge research and development investments of companies making up the sample, and having their patents generally a 20 years duration, it is possible to hypothesize that for these companies investing additional high resources also in sustainable activities requires a greater availability of time series data than the one currently considered; so that a greater investment can be adequately captured and positively impact on firm value.

Appendix

See Tables 9.14, 9.15, 9.16, 9.17, 9.18, 9.19 and 9.20.

[36] Instead, in ROA analysis none of explanatory variables was statistically significant.

[37] For pharmaceutical sector, Datastream does not display ESG scores prior to 2004.

Table 9.14 Companies included in the sample

Company name	*Country*	*Market cap (thousands of US$)*[a]	*Observation period*
Abbott Laboratories	USA	151,574,338	2004–2018
AbbVie	USA	131,064,458	2013–2018
Abeona Therapeutics	USA	118,446	2017–2018
Akorn	USA	515,936	2015–2018
Alkermes	USA	3,309,593	2015–2018
Allergan	USA	60,242,441	2004–2018
Almirall	Spain	2,947,773	2017–2018
Amneal Pharmaceuticals A	USA	856,408	2015–2018
Amphastar Pharms	USA	812,782	2016–2018
Assertio Therapeutics	USA	61,971	2015–2018
AstraZeneca	Great Britain	122,523,269	2004–2018
Bausch Health (NYS) Companies	Canada	10,031,371	2005–2018
Bayer	Germany	72,493,897	2004–2018
Beijing Tong Ren Tang 'A'	China	5,175,410	2017–2018
BioMarin Pharm	USA	13,477,513	2009–2018
Boiron	France	624,303	2018
Bristol Myers Squibb	USA	95,443,110	2004–2018
Cambrex	USA	2,017,729	2015–2018
Celltrion Pharm	South Korea	1,138,004	2018
China Med.Sy.Hdg.(Di)	Cayman	3,809,088	2014–2018
China Res.Sanjiu Med.& Pharm.'A'	China	3,957,955	2017–2018
China Resources Pharmaceutical Group	Hong Kong	5,548,098	2018
China Shineway Pharm.Gp	Hong Kong	868,505	2010–2018
China Traditional Chi Medicine	Hong Kong	2,316,146	2018
Chugai Pharm	Japan	46,225,891	2004–2018
Corcept Therapeutics	USA	1,875,968	2016–2018
Correvio Pharma	Canada	114,648	2008–2018
CSPC Pharmaceutical Gp	Hong Kong	16,572,496	2014–2018
Dong-E-E-Jiao 'A'	China	3,278,887	2017–2018

(continued)

Table 9.14 (continued)

Company name	*Country*	*Market cap (thousands of US$)*[a]	*Observation period*
Eli Lilly	USA	108,727,597	2004–2018
Endo International	USA	1,027,237	2008–2018
Faes Farma	Spain	1,510,281	2005–2018
Gilead Sciences	USA	82,378,620	2006–2018
GlaxoSmithKline	Great Britain	109,276,223	2004–2018
GUANGZHOU BAIYUNSHAN Pharm.Hdg.'A'	China	7,781,976	2017–2018
Guerbet	France	717,084	2018
H Lundbeck	Denmark	6,929,059	2004–2018
Hanmi Pharm	South Korea	3,197,530	2015–2018
Heska	USA	721,511	2016–2018
Hikma Pharmaceuticals	Great Britain	5,877,697	2006–2018
Horizon Therapeutics Public	USA	5,652,278	2015–2018
Huadong Medicine 'A'	China	6,251,119	2017–2018
Hutchison China MediTech	Hong Kong	3,092,289	2018
Hypermarcas On	Brazil	5,194,128	2010–2018
Indivior	Great Britain	389,287	2014–2018
Innoviva	USA	1,259,911	2016–2018
Insys Therapeutics	USA	6,115	2017–2018
Ipsen	France	8,967,234	2008–2018
Jazz Pharmaceuticals	USA	7,854,776	2014–2018
Jiangsu Hengrui Medicine 'A'	China	58,493,969	2017–2018
Johnson & Johnson	USA	355,021,877	2004–2008
Jointown Pharm.Gp.'A'	China	3,514,669	2018
Kalbe Farma	Indonesia	5,032,560	2011–2018
Kangmei Pharm.'A'	China	2,420,995	2018
Knight Therapeutics	Canada	785,353	2017–2018
KOBAYASHI Pharmaceutical	Japan	6,191,460	2018
Krka	Slovenia	2,349,311	2017–2018
La Jolla Pharm	USA	163,427	2016–2018
Luye Pharma Group	Hong Kong	2,590,227	2015–2018

(continued)

Table 9.14 (continued)

Company name	*Country*	*Market cap (thousands of US$)*[a]	*Observation period*
Medicines Company	USA	4,138,026	2015–2018
Merck & Company	USA	216,570,445	2004–2018
Merck KGaA	Germany	51,884,983	2004–2018
Mylan	USA	9,135,703	2005–2018
Novartis 'R'	Switzerland	203,104,422	2004–2018
Novo Nordisk 'B'	Denmark	135,161,233	2004–2018
Orion B	Finland	6,201,190	2004–2018
Otsuka Holdings	Japan	23,732,187	2012–2018
Pacira BioSciences	USA	1,789,778	2015–2018
Paratek Pharmaceuticals	USA	110,220	2016–2018
PetIQ Cl.A	USA	678,781	2017–2018
Pfizer	USA	206,312,082	2004–2018
PharmaMar	Spain	446,213	2004–2018
Radius Health	USA	1,107,925	2015–2018
RECORDATI INDUA.CHIMICA	Italy	8.603.198	2015–2018
Richter Gedeon	Hungary	3,556,791	2011 - 2018
Roche Holding	Switzerland	256,255,155	2004–2018
Sanofi	France	114,934,729	2004–2018
Shai.Fosun Pharm.(GROUP) 'A'	China	9,894,455	2014–2018
Shanghai Pharm Hdg.'A'	China	7,290,545	2012–2018
Shdg.Buchang Pharms.'A'	China	3,149,892	2018
Shenzhen Salubris Pharms.'A'	China	2,789,204	2017–2018
Sichuan Kelun Pharm.'A'	China	4,960,755	2017–2018
Siegfried 'R'	Switzerland	1,708,421	2018
Sihuan Pharm.Hdg.Gp	Hong Kong	1,209,331	2011–2018
Sino Biophm	Hong Kong	18,012,756	2013–2018
Spectrum Pharms	USA	937,806	2016–2018
SSY Group	Hong Kong	2,523,444	2018
Stada Arznei N	Germany	5,930,316	2004–2018
Supernus Pharmaceuticals	USA	1,158,382	2016–2018

(continued)

Table 9.14 (continued)

Company name	*Country*	*Market cap (thousands of US$)*[a]	*Observation period*
Tasly Pharmaceutical 'A'	China	3,217,758	2017–2018
Teligent	USA	33,447	2016–2018
Teva Pharmaceutical Industries	Israel	10,537,210	2004–2018
TherapeuticsMD	USA	661,099	2015–2018
Tong Ren Tang Techs.'H'	China	1,212,516	2018
Tonghua Dongbao Pharm. 'A'	China	4,239,663	2017–2018
UCB	Belgium	15,060,987	2005–2018
Vertex Pharms	USA	53,974,197	2004–2018
Vifor Pharma	Switzerland	10,365,865	2007–2018
Yuhan	South Korea	2,195,617	2010–2018
Yunnan Baiyao Group 'A'	China	16,174,433	2018
Zhangzhou Pientzehuang Pharms.'A'	China	9,012,941	2017–2018
Zoetis A	USA	56,217,863	2013–2018
Zogenix	USA	1,997,933	2017–2018

[a]Represents the current total market value of a company based on the current price and current shares outstanding converted to US dollars using the current exchange rate. For companies with more than one type of common/ordinary share, market capitalization represents the total market value of the company

Source Thomson Reuters Datastream; authors' elaboration

Table 9.15 Correlation matrix of dependent, independent and control variables (dependent variable: ROA)

	ROA	*ESG*	*ENV*	*SOC*	*GOV*	*CTR*	*CMB*	*TA*	*DEV*	*MBV*
ROA	1									
ESG	0.239	1								
ENV	0.291	0.853	1							
SOC	0.100	0.872	0.681	1						
GOV	0.212	0.713	0.414	0.436	1					
CNT	−0.057	−0.469	−0.371	−0.457	−0.295	1				
CMB	0.185	0.781	0.679	0.652	0.582	0.136	1			
TA	0.311	0.698	0.605	0.612	0.496	−0.638	0.298	1		
DEV	−0.066	0.0635	−0.018	0.107	0.090	−0.231	−0.090	0.169	1	
MBV	−0.067	0.0459	0.019	0.078	0.006	0.075	0.092	−0.055	−0.246	1

ROA is the firm profitability, ESG is the overall score, ENV/SOC/GOV its components score, CNT is the controversial score, CMB is the combined score, TA is the total assets, DEV is the ratio between total debt and enterprise value, MBV is the market to book value

Source Thomson Reuters Datastream; authors' elaboration

Table 9.16 Correlation matrix of dependent, independent and control variables (dependent variable: Tobin's Q)

	Q	*ESG*	*ENV*	*SOC*	*GOV*	*CNT*	*CMB*	*DEV*	*R&D*	*CPX*
Q	1									
ESG	−0.040	1								
ENV	−0.011	0.853	1							
SOC	−0.037	0.872	0.681	1						
GOV	−0.062	0.713	0.414	0.436	1					
CNT	0.191	−0.469	−0.371	−0.457	−0.295	1				
CMB	0.087	0.781	0.679	0.652	0.582	0.136	1			
DEV	−0.489	0.064	−0.018	0.107	0.090	−0.231	−0.090	1		
R&D	0.257	0.233	0.119	0.313	0.122	−0.093	0.202	−0.024	1	
CPX	0.141	0.234	0.273	0.137	0.114	0.009	0.241	−0.171	−0.273	1

Q is the firm value, ESG is the overall score, ENV/SOC/GOV its components score, CNT is the controversial score, CMB is the combined score, DEV is the ratio between total debt and enterprise value, R&D is the ratio between R&D expense and net sales, CPX is the ratio between capex and total assets

Source Thomson Reuters Datastream; authors' elaboration

Table 9.17 Descriptive statistics of total assets (billions of US$)

Year	*Obs*	*Min*	*Max*	*Mean*	*Std. Dev*
2004	80	0.0047	123.29	8.63	21.54
2005	81	0.0051	116.84	8.08	20.84
2006	82	0.0051	114.48	9	22.28
2007	83	0.0057	112.83	10.08	23.65
2008	87	0.0044	109.89	9.7	22.92
2009	91	0.0009	211.62	12.21	32.22
2010	97	0.00047	193.82	12.31	31.32
2011	98	0.0014	186.8	12.51	30.84
2012	98	0.0017	185.8	13.14	31.47
2013	100	0.0006	170.55	13.49	30.98
2014	101	0.02	167.73	13.62	30.46
2015	102	0.05	165.67	15.68	33.11
2016	102	0.07	169.8	16.42	33.9
2017	102	0.07	169.94	16.92	34.3
2018	101	0.06	157.5	17.25	34.66
Mean	93.67	0.02	157.10	12.60	28.97

Source Thomson Reuters Datastream; authors' elaboration

Table 9.18 Descriptive statistics of the ratio between total debt and enterprise value

Year	*Obs*	*Min*	*Max*	*Mean*	*Std. Dev*
2004	71	0.00	0.54	0.13	0.14
2005	74	0.00	0.62	0.12	0.15
2006	77	0.00	0.72	0.11	0.14
2007	80	−0.03	0.54	0.09	0.11
2008	81	−1.01	0.93	0.17	0.23
2009	84	0.00	0.77	0.12	0.14
2010	89	0.00	0.73	0.12	0.14
2011	93	0.00	1.78	0.17	0.24
2012	94	−0.01	2.11	0.17	0.26
2013	95	0.00	0.57	0.11	0.11
2014	99	0.00	0.52	0.11	0.11
2015	100	0.00	0.48	0.12	0.12
2016	102	0.00	0.87	0.15	0.16
2017	103	0.00	0.95	0.15	0.18
2018	102	0.00	1.59	0.21	0.3
Mean	89.60	0.00	0.91	0.14	0.17

Source Thomson Reuters Datastream; authors' elaboration

Table 9.19 Descriptive statistics of the ratio between R&D expense and net sales

Year	*Obs*	*Min*	*Max*	*Mean*	*Std. Dev*
2004	61	0.00	283.52	5.67	36.38
2005	64	0.002	58.66	1.72	7.86
2006	64	0.002	1,927.20	31.6	240.87
2007	66	0.00	45.65	1.46	6.11
2008	67	0.00	67.71	2.39	10.33
2009	73	0.002	8.99	0.53	1.52
2010	79	0.0008	7.45	0.43	1.27
2011	83	0.00	4.23	0.24	0.58
2012	91	0.00	15.89	0.39	1.76
2013	93	0.0005	26.31	0.49	2.75
2014	96	0.0009	2.88	0.20	0.45
2015	98	0.0008	27.52	0.52	2.83
2016	99	0.00	2,877.90	30.38	289.29
2017	100	0.00	20.3	0.52	2.21
2018	99	0.00	21.1	0.64	2.72
Mean	82.20	0.00	359.69	5.15	40.46

Source Thomson Reuters Datastream; authors' elaboration

Table 9.20 Descriptive statistics of the ratio between capex and total assets

Year	*Obs*	*Min*	*Max*	*Mean*	*Std. Dev*
2004	79	0.00	0.52	0.06	0.07
2005	82	0.00	0.12	0.04	0.03
2006	83	0.00	0.22	0.04	0.04
2007	84	0.00	0.12	0.04	0.03
2008	88	0.0003	0.24	0.04	0.04
2009	92	0.0005	0.49	0.04	0.06
2010	98	0.00	0.1	0.03	0.03
2011	99	0.00	0.19	0.03	0.03
2012	99	0.00	0.2	0.04	0.04
2013	101	0.00	0.15	0.03	0.03
2014	101	0.00	0.16	0.03	0.03
2015	103	0.00	0.13	0.03	0.03
2016	103	0.00	0.13	0.03	0.02
2017	103	0.00	0.21	0.03	0.03
2018	102	0.00	0.14	0.03	0.03
Mean	94.47	0.00	0.21	0.04	0.04

Source Thomson Reuters Datastream; authors' elaboration

References

Bird, R., Hall, A. D., Momentè, F., & Reggiani, F. (2007). What corporate social responsibility activities are valued by the market? *Journal of Business Ethics, 76*(2), 189–206.

Brown, L. D., & Caylor, M. L. (2004). Corporate governance and expected stock returns: Evidence from Germany. *European Financial Management*, 32.

Caulkin S. (2002). Good thinking, bad practice. *The Observer*, p. 7.

Cavaco, S., & Crifo, P. (2014). CSR and financial performance: Complementarity between environmental, social and business behaviours. *Applied Economics, 46*(27), 3323–3338.

Cho, S. Y., Lee, C., & Pfeiffer, R. J. (2013). Corporate social responsibility performance and information asymmetry. *Journal of Accounting and Public Policy, 32*(1), 71–83.

Chung, K. H., & Pruitt, S. W. (1994). A simple approximation of Tobin's Q. *Financial Management, 23*, 70.

Derwall, J. (2007). *The economic virtues of SRI and CSR*. Erasmus Research Institute of Management (ERIM).

Dowell, G., Hart, S., & Yeung, B. (2000). Do corporate global environmental standards create or destroy market value? *Management Science, 46*(8), 1059–1074.

Du, S., Bhattacharya, C. B., & Sen, S. (2010). Maximizing business returns to corporate social responsibility (CSR): The role of CSR communication. *International Journal of Management Reviews, 12*(1), 8–19.

Ferrell, A., Liang, H., & Renneboog, L. (2016). Socially responsible firms. *Journal of Financial Economics, 122*(3), 585–606.

Freeman, R. E. (1984). *Strategic management: A stakeholder approach* (p. 25). Prentice Hall.

Friedman, M, (1970, September 13). The social responsibility of business is to increase its profits. *New York Times*.

Godrey, P., Merill, C. B., & Hansen, J. (2009). The relationship between corporate social responsibility and shareholder value: An empirical test of the risk management hypothesis. *Strategic Management Journal, 30*(4), 425–445.

Greene, W. (2000). *Econometric analysis*. Wiley-Blackwell.

Hemingway, C. A., & Maclagan, P. W. (2004). Managers' personal values as drivers of corporate social responsibility. *Journal of Business Ethics, 50*(1), 33–44.

Kaplan, S. N., & Zingales, L. (1997). Do investment-cash flow sensitivities provide useful measures of financing constraints? *Quarterly Journal of Economics, 112*, 169–177.

Kim, M., & Kim, Y. (2014). Corporate social responsibility and shareholder value of restaurant firms. *International Journal of Hospitality Management, 40*, 120–129.

Kitzmueller, M., & Shimshack, J. (2012). Economic perspectives on corporate social responsibility. *Journal of Economic Literature, 50*(1), 51–84.

Lev, B., & Thiagarajan, S. R. (1993). Fundamental information analysis. *Journal of Accounting Research, 31*(2), 190–215.

McWilliams, A., & Siegel, D. (2000). Corporate social responsibility and financial performance: Correlation or misspecification? *Strategic Management Journal, 21*(5), 603–609.

Mehran, H. (1995). Executive compensation structure, ownership, and firm performance. *Journal of Financial Economics, 38*(2), 163–184.

Morck, R., Shleifer, A., & Vishny, R. W. (1988). Management ownership and market valuation: An empirical analysis. *Journal of Financial Economics, 20*, 293–315.

Perfect, S, B„ & Wiles, K, (1994). Alternative constructions of Tobin's Q: An empirical comparison. *Journal of Empirical Finance, 1*, 313–341.

Preston, L. E., & O'bannon, D. P. (1997). The corporate social-financial performance relationship: A typology and analysis. *Business Society, 36*(4), 419–429.

Prior, D., Surroca, J., & Tribó, J. A. (2008). Are socially responsible managers really ethical? Exploring the relationship between earnings management and corporate social responsibility. *Corporate Governance: An International Review, 16*(3), 160–177.

Servaes, H., & Tamayo, A. (2013). The impact of corporate social responsibility on firm value: The role of customer awareness. *Management Science, 59*(5), 1045–1061.

Surroca, J., Tribò, J. A., & Waddock, S. (2010). Corporate responsibility and financial performance: The role of intangible resources. *Strategic Management Journal, 31*(5), 463–490.

Takatsuki, Y. (2018). *ESG viewpoint*. BMO Global Asset Management.

Ullmann, A. A. (1985). Data in search of a theory: A critical examination of the relationships among social performance, social disclosure, and economic performance of US firms. *Academy of Management Review, 10*(3), 540–557.

Vanhamme, J., & Grobben, B. (2009). Too good to be true! The effectiveness of CSR history in countering negative publicity. *Journal of Business Ethics, 85*(2), 273–283.

CHAPTER 10

Corporate Green Bond: Issuance and Equity Market Reaction

Rachele Camacci

Green Bond—An Innovative Financial Instrument

In the last decades, the issue of global warming has been the main focus of many international discussions representing an emergency to be tackled in a resilient manner. The progressive scarcity of natural resources together with climate change consequences has globally redefined ethical values in investment decisions.

The financial system is an important enabler for the transition to a more sustainable economy. Indeed, market surveys have revealed a growing interest in Sustainable and Responsible Investments (SRI), a type of investment generating both a positive environmental and social impact and a financial return.

R. Camacci (✉)
Research Centre of Applied Economics, Catholic University of the Sacred Heart, Milan, Italy
e-mail: rachele.camacci@unicatt.it

© The Author(s), under exclusive license to Springer Nature Switzerland AG 2022

C. B. Pellegrini et al. (eds.), *Climate Change Adaptation, Governance and New Issues of Value*, Palgrave Studies in Impact Finance,
https://doi.org/10.1007/978-3-030-90115-8_10

In the Responsible and Sustainable Investments framework, green bonds are one of the most innovative tools to finance projects with a positive impact on the environment.

Although there is still some uncertainty regarding the definition of green bond, the generally accepted one has been issued by the International Capital Market Association (ICMA) according to which "*Green Bonds are any type of bond instrument where the proceeds will be exclusively applied to finance or re-finance, in part or in full, new and/or existing eligible Green Projects*" (ICMA, 2018).[1] In other words, the purpose of green bonds is financing sustainable projects, such as projects related to renewable energy, energy efficiency, pollution prevention and control, environmentally sustainable management of living natural resources and land use, terrestrial and aquatic biodiversity conservation, clean transportation, sustainable water and wastewater management, climate change adaptation, eco-efficient and/or circular economy adapted products, production technologies and processes, and green buildings (ibid.).

The current regulatory framework about the green bond is poor and mainly based on principles adopted on a strictly voluntary basis. These standards have been set to enhance comparability, transparency, and integrity in the overall bond market. The main guidelines for green bond issuance are the Green Bond Principles (GBP) first published in 2014 by ICMA, and consequently updated every two years.[2] According to the GBP, a bond must meet the following four core principles to obtain the green label:

1. Use of proceeds
2. Project Evaluation and Selection Process
3. Revenue management
4. Reporting activities

The first principle, Use of Proceeds, states that the issuer should define in the legal documentation the green projects to be financed with the

[1] ICMA, Green Bond Principles-Voluntary Process Guidelines for Issuing Green Bonds-June 2018.

[2] The Green Bond Principles Report 2018 is the updated version.

proceeds of the green bond issued. ICMA provides a list of eligible green projects.[3]

The second principle leads the issuer in the evaluation and selection process of green projects, which will be funded or refinanced. The issuer should clarify how the bond issuance meets its environmental purpose and broader strategy. Additionally, the evaluation can be integrated by an external review to provide a high degree of transparency.

The third principle concerns the tracking of the instrument. The green bond returns should be credited to a sub-account or in some way tracked by the issuer, who must also ensure that this is documented and periodically updated until the capital raised has been fully allocated to the financed project.

Finally, according to the fourth principle, issuers should publish on an annual basis a report listing the projects being financed, the amounts allocated to each and a measure of the expected impact on the environment. This documentation should be updated annually until full-allocation of proceeds and in case of relevant developments.

As mentioned above, according to ICMA recommendation, the issuers should nominate one or more independent external review providers to verify the alignment of their bond with the above four principles. For the purposes of this regulation, non-financial rating agencies have been established as consulting companies specialized in Second Part Opinion and Certifications.

It is important to mention the progressive commitment by the European Union in this content based on an elaboration of European green bond standards and certifications. The overall purpose is to make this tool more accessible and increase the transparency about "Use of Proceeds", either for the issuer or for the investor. This greater focus on the regulation also arises from the need to combat the phenomenon of *greenwashing*, a speculative behavior of companies that report eco-friendly corporation vision without a real commitment to the environment.

[3] ICMA, Green Bond Principles-Voluntary Process Guidelines for Issuing Green Bonds-June 2018, page. 5.

Evolution of the Green Bond Market

The green bond market developed, thanks to supranational entities in the first place. The first green bond, denominated as the "Climate Awareness Bond", was issued in 2007 by the European Investment Bank (EIB) to finance a sustainable project with an amount of about 600 million €. A year later, the World Bank issued its first green bond, a SEK 2.3 billion bond for Scandinavian investors. Until 2015 supranational institutions represented the major categories of issuers together with public development agencies.

The green bond market saw a rapid growth following the Paris Agreement signed in 2015 by 196 countries, representing one of the major legally binding international treaties on global warming in the last decade. According to this commitment, the countries started financing eco-friendly projects to keep global temperatures below 2 °C above pre-industrial levels. These efforts led to a relevant positive effect on the green bond market. Firstly, the total issued amount of green bond doubled in just a year as can be seen in Fig. 10.1. In 2015 green bond issuances achieved an amount of $44 billion globally whereas in 2016 they almost doubled reaching $81 billion. The following year, green bond market showed a new global record accounting for $155.5 billion, with a 78% growth compared to 2016. After a slow growth in 2018, the market trend recorded high growth rates in the last few years.

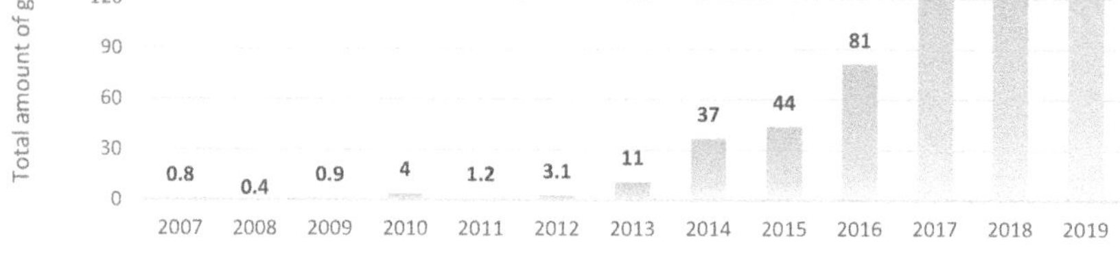

Fig. 10.1 Global Annual Green Bond Issuance 2007–2019 (*Source* Climate Bond Initiative [CBI], Moody's; personal elaboration)

Secondly, over time new issuers' categories entered the market, such as corporations with corporate green bonds, shaping a more inclusive and diversified scenario. As we can see in Fig. 10.2, in 2014 the main issuers were still the development banks, even though a wide share of the overall annual issuances already belonged to corporates including both financial and non-financial ones. The rise of corporate green bonds over time reflects the growing corporations' commitment on environmental issues. Especially, from a reputational point of view, a company issuing a green bond becomes more credible and transparent about its sustainable policy to stakeholders. Additionally, there is a clear sign of an increasing interest of governments and sovereigns towards sustainable bonds leading to sector's institutionalization.

From a geographical perspective, as showed in Fig. 10.3, over the past five years Europe has been the leading region issuing green bonds, thanks to the contribution given by France, Germany, and the Netherlands. Then we can find North America and Asia–Pacific, with their most contributing countries, respectively, being the US and China.

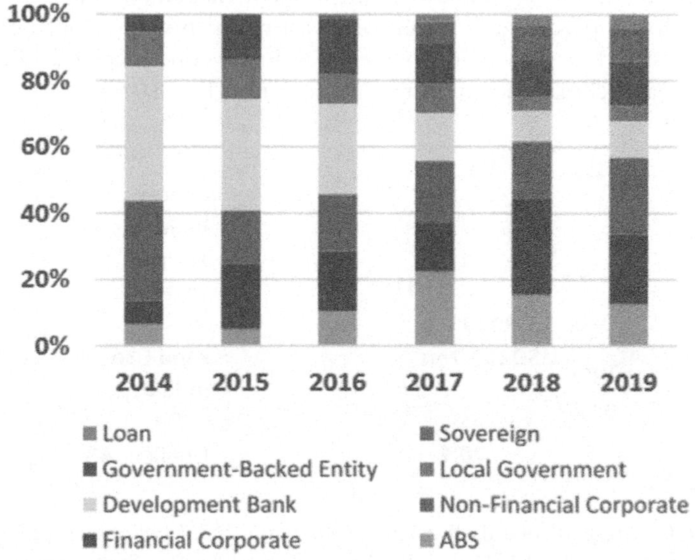

Fig. 10.2 Annual Green Bond Issuance by Issuer Type (*Source* Climate Bond Initiative [2020] "2019 Green Bond Market Summary")

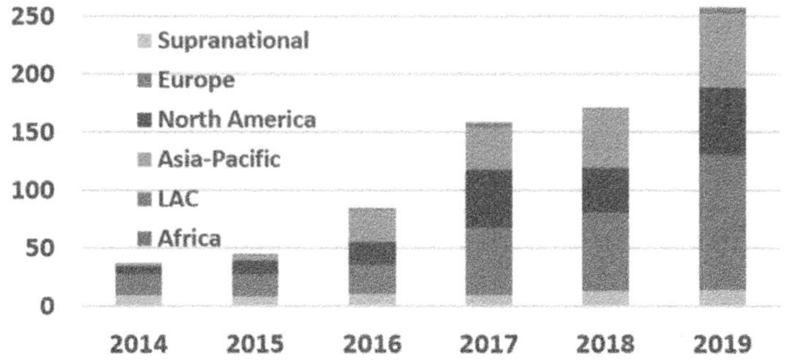

Fig. 10.3 Global Annual Green Bond Issuance by Region (*Source* Climate Bond Initiative [2020] "2019 Green Bond Market Summary")

According to CBI's analysis on the use of proceeds, the top two financed sectors are energy (31%) and buildings (30%) reflecting the recent attention to renewable energy sources and energy efficiency in the construction of new buildings. The remaining proceeds are invested in low-carbon transportation (20%) and water management (nine percent), whereas lower percentages result for investments in waste (four percent), agriculture and forestry (three percent), industry (one percent), ITC (one percent), and climate adaptation (one percent) (Fig. 10.4).

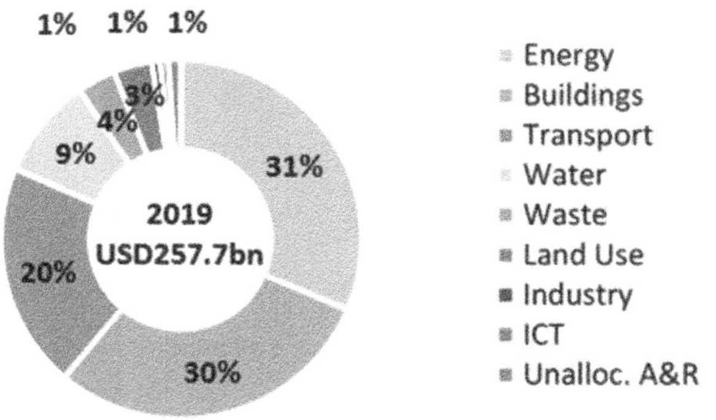

Fig. 10.4 Use of Proceeds Breakdown for USD 257.7 billion at 2019 (*Source* Climate Bond Initiative [2020] "2019 Green Bond Market Summary")

Literature Review

In the history of financial markets, the Green Bond is still a new financial instrument and there are mixed opinions on its efficiency in addressing climate change.

The academic literature is rather recent, the first studies related to the green bond instrument date back to 2013 when the green bond market first reached a significant amount of issuance.

A wide section of literature focuses on green bond pricing, also named "Greenium", demonstrating that factors unrelated to credit risk as green label could influence bond yields spread. Ehlers and Packer (2017) compare green bonds with conventional bonds finding a small bond market premium for green bonds and an average spread difference of around 18 basis points. In other words, investors' preference for green bonds leads to a higher bond price and thus a lower cost of debt. Other studies such as Zerbib (2019) and Barclays (2015) have found similar results. The reason for this effect on green bond pricing is due to a higher demand than supply, as investors are increasingly interested in green bonds and more generally in sustainable investments.

Another part of this literature documents the impact of corporate green bond issuance on the stock market, considering that the firm's performance in the bond market and the stock market are correlated. In this field, the results of studies focused on the stock market response are inconsistent. For instance, Flammer (2021) identified a positive reaction of the stock market to the announcement of green bond issuance as well as an improvement of companies' operating performance creating value. More studies, as Baulkaran (2019) and Zhou and Cui (2019) confirmed these results. Especially, Zhou and Cui investigate on Chinese green bond market confirming the positive impact on companies' stock price, innovation capacity, and companies' corporate social responsibility (CSR) improvement. Nonetheless, other scholars as Lebelle et al. (2020) find contrary results.

In conclusion, the overall results demonstrate the awareness of investors in choosing financial instruments recognizing the added value of sustainable projects. Corporate green bonds are virtuous instruments able to generate an environmental footprint left by the company and at the same time positively contribute to its financial performance.

Aim of Research: First Empirical Evidences on Green Bond Issuances

The purpose of this research is to investigate the relationship between the announcement of a corporate's green bond issuance and the resulting effect on its returns on shares.

In this regard, two main research hypotheses have been tested with different methodologies. The first hypothesis states the positive reaction of the equity market to companies' announcements of green bond issuances and is tested with an event study approach. The second hypothesis affirms the actual correlation between the green label on bonds and stocks' returns and is verified with a regression analysis.

For the purpose of this analysis, data have been collected to create two different bond samples to be compared. A sample relates to green bonds issued by a given list of companies, whereas the other one is used as a benchmark and therefore is composed by non-green bonds issued by the same target companies.

In the conduction of the event study analysis, the date of the bond's issuance announcement is a necessary information, so the Bloomberg database has been chosen to compile the two samples. In the green bond sample, bonds labeled as "green bonds/loans" have firstly been extracted from Bloomberg's fixed income database. Subsequently, two filters have been applied to exclude Government[4] bonds and to select the European area, obtaining a sample of 810 European corporate green bonds. The following step has been removed from the sample bonds with a missing announcement date and with no financial data. Therefore, the final green bond sample used in this analysis is composed of 90 corporate green bonds issued by 52 European companies in a time horizon going from 2013 to 2019.

On the other hand, the non-green bond sample was defined to create a benchmark in the analysis. This sample has been collected among bonds issued by the same target companies in the same time horizon and is composed of 149 conventional bonds. For every bond, the following information has been downloaded from the terminal: amount, coupon,

[4] Issuers with BICS (Bloomberg Industry Classification System) equal "Government" have been excluded.

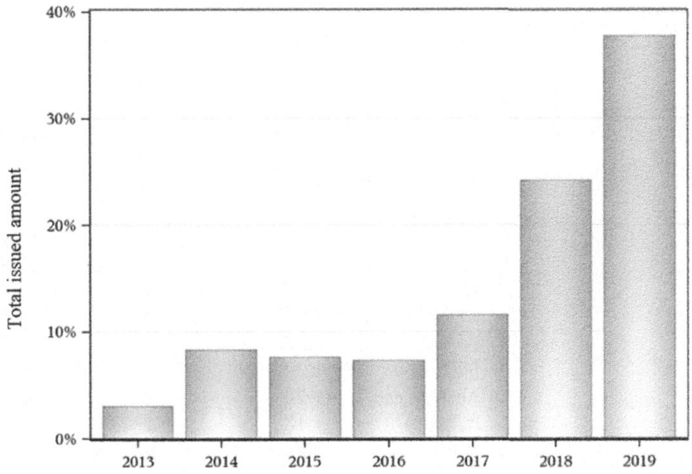

Fig. 10.5 Corporate Green Bonds Sample over time 2013–2019 This figure shows the total issuance amount (in million euros) of Corporate Green Bonds sample on an annual basis. (*Source* Bloomberg; Author's elaboration)

type of coupon,[5] announcement date, maturity, type of maturity,[6] and Bloomberg composite rating.[7]

Figure 10.5 shows the green bond sample breakdown by years of issuance. As can be observed, the major portion of Green bond has been issued in the last few years registering a rapid growth, whereas until 2016 the share of green issuances slowly increased. Therefore, the sample presents features consistent with the green bond market trend observed in the empirical evidence surveys mentioned earlier.

Table 10.1 provides the sample breakdown across by countries. For every country, the information reported are market share, number of bonds, and issued amount. It can be noticed that the top four countries for issued amount are in order France, Italy, the Netherlands, and Sweden. It is not by chance that France is the main issuer in the sample, since the

[5] Type of coupon can be fixed or variable.

[6] Type of maturity can be callable, puttable, or convertible.

[7] Bloomberg composite rating is a composed index of ratings from Moody's, S&P, Fitch, and DBRS. The rating agencies are equally weighted in the rating calculation. If the index is between two ratings, the average is rounded down to the lower rating.

Table 10.1 Corporate Green Bonds across Countries

Country	*Market Share (%)*	*# Bond*	*Amount (Million €)*
France	40.18	22	16,992
Italy	13.36	10	5,650
The Netherlands	12.8	5	5,414
Sweden	11.7	40	4,947
Spain	6.34	3	2,680
Denmark	5.98	4	2,530
Germany	3.55	2	1,500
Portugal	2.36	1	1,000
Finland	1.34	1	567
Austria	1.20	3	508
Belgium	1.18	1	500
Total	100	90	42,288

This table reported the total issuance amount (in million €) of corporate green bonds sample by country.
Source Bloomberg; Author's elaboration

first corporate green bond ever was announced in 2013 by the French company Électricité de France (EDF)[8] and according to CBI's survey,[9] it also belongs to the top issuer countries. Needless to say, thanks to the contribution of Germany and France, Europe is the main region issuing green bond worldwide being second only to the United States.

Another interesting breakdown of the sample is by industry as shown in Table 10.2. The sample is not equally distributed across different sectors; indeed, issuers operate primarily in financial and utilities sectors. The reason behind the financial sector's strong interest in the green bond market is justified by the existing negative relationship between extreme weather events caused by climate change and financial returns. These negative effects have increased climate risk awareness and consequently efforts to mitigate the global issue. Indeed, new forms of sustainability-related investments are emerging in the financial sector, such as investments to reduce greenhouse gas emissions and those that improve

[8] Électricité de France (EDF) is one of the companies in the sample analysed.

[9] Climate Bond Initiative—Green Bond Market Summary Q3 2020, page 1.

Table 10.2 Corporate Green Bonds across Industry

Industry	*Market Share (%)*	*# Bonds*	*Amount (Million €)*
Utilities	47.54	23	20,103
Financials	43.03	58	18,195
Industrial	2.91	2	1,230
Discretionary consumption	2.82	4	1,193
Materials	1.34	1	567
Essential consumer goods	1.18	1	500
Energy	1.18	1	500
Total	100	90	42,288

This table presents the total issuance amount (in million €) of corporate green bonds sample by industry, according to Bloomberg's BICS (Bloomberg Industry Classification System) codes. *Source* Bloomberg; Author's elaboration

climate resilience. It is evident that climate change is creating a business opportunity.[10]

[10] Climate Change and the Financial Sector: A Time of Risk and Opportunity—Miller A S, Swann S A (2016).

Table 10.3 Descriptive Statistics of Green Bond

Variable	*Obs*	*Mean*	*Median*	*Std Dev*	*Min*	*Max*
Amount (Million €)	90	470	500	555	3	3,709
Fixed-rate bond	90	0.58	1.00	0.49	0	1.00
Coupon (%)	52	1.48	1.15	0.96	0.13	3.77
Maturity (years)	88	8.35	5.00	13.07	2.00	100
Bloomberg composite rating	40	BBB +	A–		BB	AA

Source Bloomberg; Author's elaboration

Table 10.4 Descriptive Statistics of Issuers

Variable	*Obs*	*Mean*	*Median*	*Std Dev*	*Min*	*Max*
Total Asset (Billion €)	239	318.77	122.77	478.25	0.21	2,070
ROA (%)	239	3.63	1.37	3.94	0.02	16.48
MTBV	239	1.09	1.00	0.51	0.26	3.64
Tobin's Q	239	0.87	1.05	0.37	0.11	2.35
Leverage (%)	239	315	182	289	25	1,153

Source Thomson Reuters Datastream; Author's elaboration

The following tables show the descriptive statistics of the green bond sample.[11] In Table 10.3, the features of green bonds sample are represented such as issued amount, fixed-rate bond,[12] coupon, maturity, and Bloomberg composite rating. Moreover, in Table 10.4 the summary statistics of issuer belonging to sample are identified.

[11] The descriptive statistics of the green bond sample showed in the Table 10.3 and Table 10.4 corresponding to the variables used in the regression model.

[12] Fixed-rate bond is a dummy variable, it is equal to one if the bond is a fixed coupon payment.

A First Approach: The Event Study

The event study methodology[13] has been used to examine the market reaction to Green Bond issuance announcements to verify whether these events have an impact on stocks' value. This effect has been measured by computing abnormal returns on share prices using an economic model, namely the market model representing one of the most widely used in this type of analysis.

The present study defines as event date "0" the announcement date for each bond, as event window the time horizon including 10 days before and after the announcement date [-10; 10] and as estimation window the time range [−220; −11]. Having defined the time horizon of 210 trading days, it has been possible to collect companies' historical share prices and reference market indexes[14] from the Bloomberg database.

According to the market model, the coefficients α_i and β_i have been estimated through Ordinary Least Squares (OLS) in the estimation window [−220; −11] to compute the expected daily returns on stocks. Then by using these results abnormal returns have been computed as the difference between actual stock returns[15] and expected returns or "normal" returns. The time intervals considered are [−10, −6], [−5, −2], [−2, 2], [−2, 1], [−1, 1], [−1, 0], [2, 5], and [6, 10], besides the main window [−1, 0]. Thereafter, the Cumulative Abnormal Return (CAR) have been computed with the following formula:

$$CAR_i(\tau_1, \tau_2) = \sum_{t=\tau_1}^{\tau_2} AR_{i,t}$$

Since the purpose of the study is examining the average share price trend when the event occurred, it has been necessary to aggregate CARs by time window and securities. The mean value of CARs for the given event windows mentioned above is named Cumulative Average Abnormal

[13] Event studies in economics and finance - MacKinlay A. C. (1997); The event study methodology since 1969—Binder J. (1998). Measuring security price performance—Brown S. and Warner J. (1980). Using daily stock returns: The case of event studies—Brown S. and Warner J. (1985).

[14] The reference indices used to calculate market returns are the main market indices of the country (e.g. CAC 40 for France, IBEX 35 for Spain, and FTSE MIB for Italy).

[15] Actual stock return is calculated on share price downloaded from Bloomberg.

Return and have been obtained as follows:

$$\overline{CAR}(\tau_1, \tau_2) = \frac{1}{N}\sum_{i=1}^{n} CAR_i(\tau_1, \tau_2)$$

At this stage, the significance of the results has been tested with a parametric statistical test allowing to check whether the mean value of the distribution significantly differs from a specific reference value. Tables 10.5 and 10.6, respectively, show the event study results of green bond and conventional bond samples reporting for each time window the

Table 10.5 Event Study Results for Green Bond Sample

	Obs	Mean (%)	Std Dev	t-stat
CAR [−10; −6]	90	−0.286	0.030	−0.904
CAR [−5; −2]	90	0.164	0.022	0.701
CAR [−2; −2]	90	0.605	0.030	1.931
CAR [−2; 1]	90	0.592	0.026	2.128**
CAR [−1; 1]	90	0.306	0.021	1.388
CAR [−1; 0]	90	0.413	0.014	2.734***
CAR [2; 5]	90	0.051	0.023	−0.210
CAR [6; 10]	90	0.674	0.028	2.295**
*** p < 0.01, ** p < 0.05, * p < 0.1				

Source Bloomberg; Author's elaboration

Table 10.6 Event Study Results for Conventional Bond Sample

	Obs	Mean (%)	Std Dev	t-stat
CAR [−10; −6]	149	−0.152	0.024	−0.770
CAR [−5; −2]	149	0.154	0.022	0.810
CAR [−2; −2]	149	0.131	0.025	0.620
CAR [−2; 1]	149	−0.120	0.024	−0.644
CAR [−1; 1]	149	−0.250	0.019	−1.620
CAR [−1; 0]	149	−0.063	0.018	0.393
CAR [2; 5]	149	0.320	0.024	1.642
CAR [6; 10]	149	−0.040	0.031	−0.159
*** p < 0.01, ** p < 0.05, * p < 0.1				

Source Bloomberg; Author's elaboration

mean of Cumulative Abnormal Returns, the relative Standard Deviation, and the t-statistic.

Regarding the green bond sample, the cumulative abnormal return is on average positive in almost all event windows as can be seen in Table 10.5. The CAR for the event window [−1, 0] with a value of 0.41% reaches the one percent threshold of statistical significance. Also, in the event windows [−2; 1] and [6; 10], the CARs are positive and statistically significant at a five percent level. Overall, the results demonstrate a positive relationship between announcements of corporate green bond issuance and their stock price values. However, these findings are consistent with other empirical studies (Flammer, 2021; Roslen et al., 2017) resulting as well in significant CARs in the same two-day event window [−1; 0]. Shifting to Table 10.6, the average CARs obtained in the conventional bond sample have both positive and negative values in different time windows, even though never statistically significant.

As a general comment, there appears to be a different stock market reaction following the announcement of green bonds and conventional bonds. According to the results obtained, the issuance of conventional bonds does not generate any extraordinary behavior of investors whereas there is a positive feedback in the stock market for green bond announcements. To confirm this finding, a t-test has been applied among the two independent samples and the results are shown in Table 10.7.

As a result, in the time windows [−1; 0] and [−2; 1] there is a statistically proved difference among the two samples. Therefore, it is possible to state that the issuance of Green Bonds leads to a significant reaction on the market, as opposed to conventional bonds. Thanks to these

Table 10.7 Results of t-test on the two independent samples

	Std Dev Pooled	t-stat
CAR [−10; −6]	0.026	−0.807
CAR [−5; −2]	0.022	0.553
CAR [−2; −2]	0.027	1.665
CAR [−2; 1]	0.025	1.797*
CAR [−1; 1]	0.020	1.173
CAR [−1; 0]	0.017	1.847*
CAR [2; 5]	0.024	−0.173
CAR [6; 10]	0.030	1.697

*** $p < 0.01$, ** $p < 0.05$, * $p < 0.1$

Source: Bloomberg; Author's elaboration

obtained results from the event study analysis, the first research hypothesis mentioned earlier is confirmed to be true.

In order to answer the second research hypothesis and confirming the impact of green label on positive CAR values, a regression model is defined and applied.

Further Empirical Analysis

As previously mentioned, the purpose of this second analysis is to identify the main variables explaining the value of the cumulative abnormal returns and investigate the significance of the green label as explanatory variable in the model.

The regression model has the following form:

$$\begin{aligned} CAR_i(t_{1,}t_2) = \alpha_i + \beta_i \times Green_{label} + \beta_i \times AmountIssued + \beta_i \times Coupon \\ + \beta_i \times Maturity + \beta_i \times MaturityType + \beta_i \times Tobin'sq \\ + \beta_i \times Size + \beta_i \times ROA + \beta_i \times MTBV + \beta_i \times Leverage \end{aligned}$$

The dependent variable corresponds to the CARs value obtained in the event study analysis whereas the independent variables can be ideally grouped into two main typologies based on their characteristics.

A first typology concerns the bond's features and includes the following variables:

- Green_label: a dummy variable equal to one if the bond is green and zero otherwise. A positive and statistically significant coefficient confirms the second research hypothesis, a reason why it is the main variable of interest in this study.
- Amount Issued: the natural logarithm of the bond's issued amount.
- Coupon: the value of the bond's coupon rate.
- Maturity: the number of years to the maturity date.
- Maturity Type: a set of dummy variables specifying different types of maturities, such as Callable, Puttable, Convertible, and Perpetual/Callable.

The second group is composed of variables related to the issuing company's features,[16] as:

- Tobin's q: a proxy of growth opportunities. The index has been computed as the Market value of equity plus book value of preferred stocks and debt divided by book value of total assets.
- Size: the natural logarithm of Total Assets commonly used as a measure of firm size.
- *ROA*: a proxy of company's profitability, namely Return on Assets.
- *MTBV*: a ratio between a company's current market value and its book value, namely Market To Book Value.
- Leverage: an index measuring the level of debt, computed as a ratio between Total Debt and Common Equity.

The OLS regression model has been adopted to perform this analysis since the data are cross-section. Subsequently to verify the consistency with the OLS model's assumptions, the following statistical tests have been applied:

- Durbin-Watson test, to test autocorrelation.
- Breusch-Pagan test, to test heteroscedasticity.

A problem of heteroscedasticity arises in the model therefore the OLS estimators are unbiased, but the standard errors are incorrect. To solve this issue, the OLS model with robust standard errors for heteroscedasticity has been chosen.

As mentioned above, the dependent variable corresponds to the CARs value and for the purpose of this analysis only significant event windows have been chosen. Therefore, according to the CAR tests the study has just focused on the two event windows [−1; 0] and [−2, 1], the reason why there are two regression models. Table 10.8 shows the regression model output by using CAR as dependent variable in the event window [−1; 0].

[16] The independent variables concerning the company's features refer to the previous fiscal year from the issue date. These variables have been drawn from Thomson Reuters Datastream database.

Table 10.8 OLS Model using CAR in the event window [−1; 0]

Model: OLS, using observations 1-239
Dependent variable: CAR [− 1;0]
Heteroskedasticity-robust standard errors, variant HCO

Variable	Coefficient	Std Error	t-ratio	p-value
Const	−0.22706196	0.32851603	−0.69120	0.4904
Amount Issued	−0.00288148	0.00122657	−2.34920	0.01997**
Coupon	−0.00166947	0.00124927	−1.33640	0.18323
Maturity	0.00002	0.00232545	0.00860	0.99315
Green label	0.00460665	0.00241975	1.90380	0.05864*
Callable	0.00702019	0.00379986	1.84750	0.06642*
Convertible	0.01346555	0.01700779	0.79170	0.42963
Putable	0.02590908	0.00443587	5.84080	2.60E-08***
Perp_Cal	−0.00212596	0.00635583	−0.33450	0.73842
Tobin's Q	0.04080744	0.02345709	1.73970	0.08374*
Size	0.01048344	0.01274948	0.82230	0.41209
ROA	0.00034893	0.00082998	0.42040	0.67472
MTBV	−0.00706013	0.01059525	−0.66630	0.5061
Leverage	0.00415031	0.01360401	0.3051	0.76068
Observations	239			
R-squared	0.4			
Year FE	Yes			
Industry FE	Yes			

*** p < 0.01, ** p < 0.05, * p < 0.1

Source Thomson Reuters Datastream; Author's elaboration

As a first consideration, it is interesting to underline that the Green label variable is positive and statistically significant at ten percent level. This finding confirms the positive effect of Green label on CAR value, specifically when the bond is green there is an impact on the value of the cumulative abnormal return of about 0.4%. Additionally, according to the output the amount issued resulted to be significant at the five percent level, but this variable has a negative relationship with the CAR. When the issued amount increases by one unit, the cumulative abnormal return falls by 0.2%, *ceteris paribus*. The relationship between the bond's issued amount and the CAR value is strictly correlated to the bond's liquidity. Also, the two dummy variables Callable and Putable related to the bond's maturity type resulted statistically significant both generating a positive effect on the abnormal returns' value. In particular, when the Callable (Putable) variable equals to one, the CAR value increases by 0.7% (2.5%).

Finally, Tobin's Q index appears to be significant at the five percent level and with a positive coefficient of four percent. This result is in line with previous findings by Baulkaran (2018) emphasizing a stronger market reaction for companies with growth opportunities proxied by Tobin's Q.

Looking at the second regression model output, except for the Callable variable, all the previously highlighted variables referred to the event window [−1; 0] remain statistically significant and with the same sign also in the event window [−2; 1]. In addition, the Maturity variable becomes significant at the 5% level with a negative sign and a coefficient of 0.8%, all other things being equal. According to the latter result, when the company issues fixed income securities with a longer maturity, stakeholders may reflect a signal of uncertainty. The Convertible variable also presents a negative link with the value of the cumulative abnormal returns, the presence of this bond specification decreases the value of the CAR by two percent, whereas Puttable bonds increase the dependent variable by 3.7%. Overall, the most important result of this analysis refers to the one recorded by Green label, as can be seen in Table 10.9, the second model confirms the statistical significance and the positive effect for Green label variable.

In conclusion, according to the empirical evidence of this further analysis through the regression model, it is possible to positively answer the second research hypothesis. Therefore, it can be stated that the Green label variable explains the cumulative abnormal return value with a positive impact.

Concluding Remarks

In this historical period characterized by climate change and increasing scarcity of natural resources, the environmental sustainability is one of the most debated topics. The Paris Agreement and the more recent European Green Deal represents governments' efforts towards a more environmentally and socially conscious economy. In this context, the financial system has played a fundamental role in the transition to low-carbon economy generating a change to sustainable finance. Moreover, many sustainability-related financial instruments have been created in recent decades, attracting the interest of many stakeholders.

The chapter focused on one of the most innovative financial instruments, the Green Bond. The purpose of this analysis is to contribute to the existing literature providing empirical evidence about the efficiency

Table 10.9 OLS Model with CAR in the event window [−2; 1]

Model: OLS, using observations 1-239
Dependent variable: CAR [−2;1]
Heteroskedasticity-robust standard errors, variant HCO

Variable	Coefficient	Std Error	t-ratio	p-value
Const	0.24620675	0.53166094	0.46310	0.6439
Amount Issued	−0.00156913	0.00151978	−1.03250	0.30333
Coupon	−0.0003292	0.00117186	−0.28090	0.77912
Maturity	−0.00830031	0.00337249	−2.46120	0.01485**
Green label	0.00695411	0.00405664	1.71430	0.08832*
Callable	0.00498526	0.00508069	0.98120	0.32789
Convertible	−0.02035381	0.00860893	−2.36430	0.0192**
Putable	0.03716256	0.00657859	5.64900	6.71E-08***
Perp_Cal	0.00639154	0.00941828	0.67860	0.4983
Tobin's Q	0.08034353	0.03684817	2.18040	0.03061**
Size	−0.01174309	0.02187005	−0.53690	0.59201
ROA	0.00053937	0.00125373	0.43020	0.66759
MTBV	−0.02654514	0.01667343	−1.59360	0.12218
Leverage	0.00570574	0.02282237	0.25000	0.80289
Observations	239			
R-squared	0.3661			
Year FE	Yes			
Industry FE	Yes			
*** p < 0.01, ** p < 0.05, * p < 0.1				

Source Thomson Reuters Datastream; Author's elaboration

of green bonds. Specifically, the goal of this study is to understand how the stock market reacts to the announcement of Corporate Green Bond issuances in a European context. This analysis mainly takes inspiration from two previous studies. The first one is the analysis conducted by Flammer (2013, 2021), in which a positive reaction to the announcement of green bond issuances and an improvement in the financial and operational performance of the issuer have been found. The second research study refers to Baulkaran (2018) resulting in a positive stock market reaction to green bond issuance announcements and in a positive relationship between cumulative abnormal return and firm size, Tobin's q and firm growth index.

Following the studies mentioned above, this analysis focused on the relationship between announcements of corporate green bond and the stock price values. Two bond samples have been collected in order to compare green bonds with conventional bonds issued in the period between 2013 and 2019 by European companies. Using the event study methodology and choosing the market model to estimate "normal" returns, the cumulative abnormal returns in different time windows have been obtained. The average CARs of the green bonds sample in the main event window [−1; 0] are positive and statistically significant at the one percent level, whereas the average CARs of conventional bonds sample are not statistically significant in any event window. Thanks to the t-test statistic computed on two independent samples, two event windows [−1; 0] and [−2; 1] have been found statically significant underlying a difference in stock market reaction between green bond and conventional bond announcements.

In order to verify the effect of the green label on the cumulative abnormal return values, a second analysis has been carried out using a regression model. Specifically, two independent regressions have been calculated using the cumulative abnormal returns in the event windows [−1; 0] and [−2; 1] as the dependent variable. The results obtained argue a positive relationship between the green label variable and the cumulative abnormal return value. In addition, it appears to be a positive connection between Tobin's q and CARs value, whereas the Maturity variable and the amount issued variable are negatively correlated with CARs.

Overall, the empirical results explain in this chapter confirm those found by other scholars and the theory about the positive stock market reaction to green bond announcements. The statistically not significant results obtained for conventional bonds can be considered as a further confirmation of the actual recognition by stakeholders of the green bonds' efficiency.

The positive effect of the green label can be explained through several factors. Firstly, the green bond issuance is strictly correlated with a higher level of CSR transparency. Indeed, when a company decides to finance itself by issuing green bonds, it must align with the Green Bond Principles and this means to disclose all the information related to the project to be financed. ICMA recommends this information to be evaluated by auditing or certification companies which provide the green label only if the issuing company meets all the requirements.[17]

[17] For instance, CBI is an international organization defining the Climate Bond Standard and providing a Certification.

As a second consideration, green bonds have a lower financing cost due to a lower yield than conventional bonds. The reason for a lower yield in green bonds is given by the higher investors' demand than supply, in both the primary and secondary markets. A possible future reduction in the evaluation time needed to obtain the green label and in the related costs as well, could lead to an increase in the expected supply of green bonds resulting in a lower advantage for the issuer.

Another factor playing an important role in the positive effect of green label is the attractiveness of ESG investors, a segment of stakeholders more interested in sustainable investments. On the other hand, traditional investors are also interested in green bonds because they are less risky. Overall, the results of this analysis suggest that green corporate bond issuance is the best corporate choice to attract different types of investors with different types of resources.

In conclusion, the results confirm that green bonds not only generate a positive impact on the environment complying with Sustainable Development Goal 13[18] but also have a positive effect on the financial performance since the announcement of corporate green bonds is recognized as a positive shareholder value creation event.

References

Paris, A. (2015). *United Nations framework convention on climate change*. France.

Barclays. (2015). The cost of being green. US Credit Focus, September.

Baulkaran, V. (2019). Stock market reaction to green bond issuance. *Journal of Asset Management, 20*(5), 331–340.

Binder, J. (1998). The event study methodology since 1969. *Review of Quantitative Finance and Accounting, 11*(2), 111–137.

Brown, S., & Warner, J. (1980). Measuring security price performance. *Journal of Financial Economics, 8*(3), 205–258.

Brown, S., & Warner, J. (1985). Using daily stock returns: The case of event studies. *Journal of Financial Economics, 14*(1), 3–31.

Climate Bonds Initiative. (2018). The climate bonds taxonomy. A guide to climate aligned assets & projects 2018.

Climate Bonds Initiative. (2019). Green Bond European Investor Survey.

Climate Bonds Initiative. (2019). Green Bond Market Highlights 2018.

[18] Sustainable Development Goal 13 (SDG 13) belongs to the framework of the 17 Sustainable Development Goals defined by the United Nations in 2015. According to this initiative, the target will be reached in 2030.

Climate Bonds Initiative. (2019). Green bond pricing in the primary market: January-June 2019.
Climate Bonds Initiative. (2019). Green bonds market summary - Q3.
Climate Bonds Initiative. (2019). Green bonds: The state of the market 2018.
Ehlers, T., & Packer, F. (2017). Green bond finance and certification. *BIS Quarterly Review September.*
Flammer, C. (2013). Corporate social responsibility and shareholder reaction: The environmental awareness of investors. *Academy of Management Journal, 56*(3), 758–781.
Flammer, C. (2021). Corporate green bonds. *Journal of Financial Economics.*
Glavas, D. (2018). How do stock prices react to green bond issuance announcements?. Available at SSRN 3279069.
ICMA. (2018). Green bond principles voluntary process guidelines for issuing green bonds. June 2018.
Lebelle, M., Lajili Jarjir, S., & Sassi, S. (2020). Corporate green bond issuances: An international evidence. *Journal of Risk and Financial Management, 13*(2), 25.
MacKinlay, A. C. (1997). Event studies in economics and finance. *Journal of Economic Literature, 35*(1), 13–39.
Miller, A. S., & Swann, S. A. (2016). Climate change and the financial sector: A time of risk and opportunity. *Georgetown Environmental Law Review, 29*, 69.
Roslen, S. N. M., Yee, L. S., & Ibrahim, S. A. B. (2017). Green bond and shareholders' wealth: A multi-country event study. *International Journal of Globalisation and Small Business, 9*(1), 61–69.
Wang, J., Chen, X., Li, X., Yu, J., & Zhong, R. (2020). The market reaction to green bond issuance: Evidence from China. *Pacific-Basin Finance Journal, 60*, 101294.
Zerbib, O. D. (2019). The effect of pro-environmental preferences on bond prices: Evidence from green bonds. *Journal of Banking & Finance, 98*, 39–60.
Zhou, X., & Cui, Y. (2019). Green bonds, corporate performance, and corporate social responsibility. *Sustainability, 11*(23), 6881.

CHAPTER 11

The Sustainability Challenge for Asian Emerging Markets: Some Empirical Evidences

Claudia Cannas

Introduction

The term "emerging market economy" was first used in 1981 by Antoine W. Van Agtmael of the International Finance Corporation of the World Bank. Emerging markets are economies that have not already achieved a complete development, but grow at a fast pace, reducing the gap with developed economies.

The path to sustainability has been slower in emerging markets than in developed markets. The intense protests against economic globalization and accusations pointed towards companies' action of the twentieth

C. Cannas (✉)
Research Centre of Applied Economics, Catholic University of the Sacred Heart, Milan, Italy
e-mail: claudia.cannas@unicatt.it

© The Author(s), under exclusive license to Springer Nature Switzerland AG 2022

C. B. Pellegrini et al. (eds.), *Climate Change Adaptation, Governance and New Issues of Value*, Palgrave Studies in Impact Finance,
https://doi.org/10.1007/978-3-030-90115-8_11

century have highlighted the existence of a real sustainability challenge for emerging markets.

Despite developed countries, emerging markets continuously face considerable challenges. The awareness of growing issues at ESG level has inevitably led these markets to fight against several and peculiar issues of an emerging economy including poverty, urbanization, pollution, disease, and corruption. Additionally, these countries stand out for their underdeveloped, illiquid equity markets and nationalized banks, weak monitoring by bureaucrat, limited dissemination of companies' information which does not attract the attention of investor and financial analysts; eventually, in some cases for low ESG investments due the deprived social environment (Garcia et al., 2017).

Embracing the concept "Triple Bottom Line" of Elkington (1997) it's possible to analyse these challenges considering three dimensions of sustainability individually.

a) Environmental Issues

Asian countries are one of the areas most affected by the frequency and severity of extreme weather events such as droughts, floods, and typhoons. In this context, emerging countries are among the most exposed to the impacts of climate change, including problems of local pollution and resource scarcity. The pollution and policies implemented to reduce the emission by the government seem to have a strong impact in these countries. The finding of primary resources is tightly connected to the high growth rate that characterizes emerging countries.

According to the United Nations, the world's population will increase from 7.1 billion to 9.6 billion by 2050, with most of the increase in the world's poorest countries. The increase in incomes in the emerging markets leads from year to year to a strong increase in the demand for energy resources, food, and water. In this context, many emerging markets will be among the most vulnerable to water stress related to the water crises that will characterize the next decade (UN report, 2019).

Therefore, the technological progress, which is at the base of the transformation of different industries can represent a relevant key to face environmental issues determining critical consequences for global economies at the regional, national and local.

b) Social Issues

The weakness of institutions and the inequality that characterizes the regulatory regimes of these countries represent the heart of social issues. Emerging markets are characterized by several social problems due in particular to weak political institutions not always capable of effectively responding to issues that involve the community and the entire country. Moreover, there is limited consumer protection and lack of regulatory supervision of quality control or product standards.

With regard to workplace, workers often do not have any guarantee that protects them from human rights violations and exploitation, let alone sufficient health and safety standards. The working environment is generally characterized by poor transparency, and in many emerging countries slavery seems to be a current problem.

In this context, the relationship between an enterprise and the communities is often influenced. Firms in carrying out their primary function inevitably impact the surrounding environment and this consequently has an impact on the community itself that is in that specific environment. It is often possible to observe that firms' action is limited and affected by community relations.

c) Governance Issues

In Emerging Markets, the Governance standards are really different from those in developed markets. Emerging countries are strongly characterized by huge issues such as corruption, the low level of disclosure, and lack of transparency that highly affect the dimension of the governance.

Corruption tends to be strongly correlated with a country's per capita income (Odell & Ali, 2016). Indeed, emerging and frontier countries often show the lowest scores of Transparency International's corruption perception index. Despite the significant progress made in countries that have moved away from local GAAP standards towards IFRS, information is often limited. The quality and integrity of financial statements vary greatly from one company to another and from one country to another.

As regards ownership and alignment of interests, the incidence of controlling shareholders is much higher than in most developed markets. Generally, Board is mainly composed of majority shareholders; this often serves as a signal that boards are not independent. Even with independence, the requirements and experience of the management may be limited. Consequently, transactions with related parties, transfer pricing, issues of executive compensation, poor disclosure, and the lack of board independence may create a mismatch of interests between majority and minority shareholders failing to provide appropriate protection for the latter (Jensen & Meckling, 1976).

Academic Literature

The literature on the relation between Corporate Financial Performance and Environmental, Social and Governance (ESG) practices historically focused on developed markets (Bassen et al., 2006; Cheng et al., 2014; Dhaliwal et al., 2011). Emerging markets, despite representing a significant portion of business worldwide, have not yet found an expressive presence in the literature addressing ESG performance. (Baughn et al., 2007; Dobers & Halme, 2009). In part, this is because reliable data were not available until quite recently. Nevertheless, several studies highlight that sustainability does not seem to impede the development of these countries, they actually suggest a positive impact of improving ESG practices on Corporate Financial Performance.

In this chapter, only literature specific to emerging markets has been explored. The main details referring to the broad literature on the relationship between ESG factors and firm performance can be found earlier in Chapter 6.

Despite many challenges that emerging countries face, several studies show that sustainability does not seem to set limits to development in these markets, but creates new opportunities.

Friede et al. (2015) analyze that the link between ESG and corporate financial performance is even more apparent for emerging markets suggesting a major number of positive evidences for the 65,4% until 70,8% if it does not consider portfolio studies.

One of the first large-scale studies[1] analyzing the "business case for sustainability" in emerging markets shows that greater sustainability efforts, such as better governance, increased environmental practices, as well as greater investments in social and economic development, enable

[1] The study mentionated is Developing value (Thorpe J and Prakash-Mani K 2003).

these markets to achieve greater benefits, such as higher sales, improved productivity, reduced risk, and increased market access. This business case constantly develops and it is strongly influenced by the political, social, and economic context in which companies operate. In this context, governments should be primarily responsible for creating and ensuring conditions that allow companies to guarantee maximum stability and the right incentives. Moreover, the growing number of stakeholders and access to information causes a greater demand for accountability and transparency as well as an increase in the company's expectations and the contribution that the same company can make to achieving sustainable development.

Empirical evidence in literature shows that while developed markets get more benefits in terms of reputation and brand value, emerging markets take more advantage in terms of cost reduction and higher sales. In particular, the most important action areas of sustainability for emerging markets seem to be the improvement of environmental processes and practices and managing human resources.

Another critical issue for companies operating in emerging countries is the role of disclosure. The academic literature points to a positive relationship between financial performance the level of disclosure in environments heavily characterized by asymmetric information such as emerging markets. Lang et al. (2004), suggest that the positive assessment of the impact of ESG disclosure on firm performance is due to the fact that investors seem to tend to assess positively an increase in disclosure by firms for which there is a relevant lack of information. In order to solve asymmetric information problems, investors decide to invest their resources in clustered firms. Farooq and El Ouadrhiri (2013), observe that firms in emerging markets are usually clustered around financial centres. The higher density of firms in the financial centre seems to translate into a reduction of information asymmetries for firms, and consequently allows for more relevant information. In contrast, the study shows that India, one of the most important emerging markets in the World, shows a negative link between Firms' Performance with headquarters in financial centres and ESG disclosure. This negative relationship seems to be due to the fact that socially responsible activities are considered as unnecessary costs by stock market participants. Unlike to previous studies the results show additionally that disclosure has not a significant impact on firm performance in these environments, because it is less reliable than disclosure provided by companies operating in environments with less asymmetric information; therefore, it is highly likely that it is not valued by stock market participants.

Asian Firms and Esg Issues: A First Empirical Approach

The aim of this research work is to verify the existence of a positive link between financial performance and non-financial performance such as ESG (Environmental, Social and Governance) factors for Asian companies operating in emerging markets between 2008 and 2019. In order to assess this relationship, I built a sample of panel data consisting of 415 Asian companies. These companies are mainly distributed in China (103), Taiwan (85), Korea (74), and India (53). Figure 11.1 illustrates the geographical distribution of Emerging Markets companies.

In order to investigate which were the best sectors in terms of ESG development practices and policies, all sectors of *Global Industry Classification Standard* (GICS) were considered.

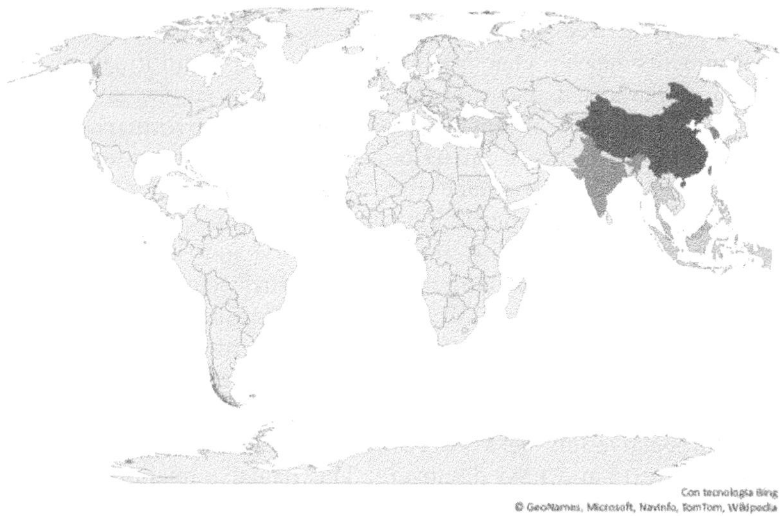

Fig. 11.1 Geographical distribution of the Sample (*Source* Thomson Reuters Refinitiv; Author's elaboration)

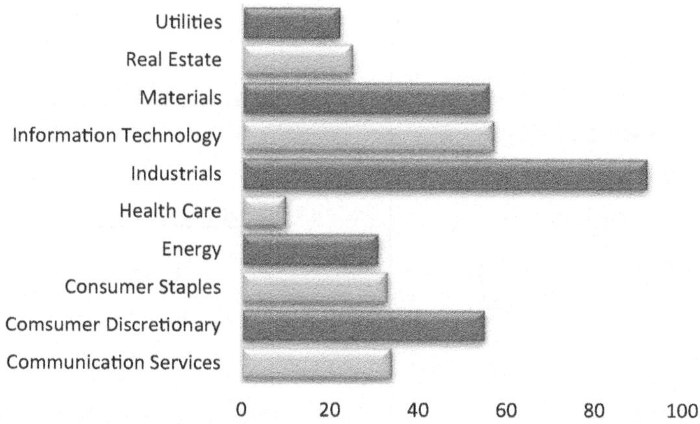

Fig. 11.2 Industry distribution of the Sample (*Source* Thomson Reuters Refinitiv; Author's elaboration)

To ensure better comparability of data, Financial companies were not involved in the sample. Generally, the empirical analyses based on Financial companies consider different variables to capture size and efficiency indicators, which could have made the analysis biased.

In Asian Emerging Markets the firms more aware of ESG issues belong to Industrial (92), Information Technology (57), Materials (56) and Consumer Discretionary sectors. Figure 11.2 shows the industry distribution of Emerging Markets companies.

In order to examine how ESG factors impact on Financial Performance, we considered accounting variables and ESG scores provided by Thomson Reuters Refinitiv. The ESG risk methodology of Refinitiv offers more than 400 different ESG metrics guaranteed by a careful standardization process and updated on an ongoing basis. This methodology includes three overall scores. It also allows studying in-depth specific subcategories. In particular, it includes 10 subcategories: ESG Score, ESG Controversies Score and ESG Combined Score,[2] providing the individual score for the ESG factors such as

[2] The ESG Combined Score considers the ESG Score in light of the controversy issues in which companies are involved.

Environmental (E), Social (Social) and Governance (G) Pillars scores:

- Resource Use Score
- Emissions Score
- Environmental Innovation Score

Environmental Pillar Score

- Workforce Score
- Human Rights Score
- Community Score
- Product Responsibility Score

Social Pillar Score

- Management Score
- Shareholders Score
- CSR Strategy Score

Governance Pillar Score

Instead, the accounting variables have been extracted from Datastream database. With regard to dependent variables, the Firm Performances can be measured by two different proxies:

- **Return on Asset (ROA)**, computed as the ratio between Earnings Before Interest, Tax and Depreciation (EBITDA) and Total Assets;
- **Tobin's Q**, computed as the ratio between the market value of equity plus the book value of the asset minus the book value of equity and the deferred taxes all on book value of the asset.

In the light of the aforementioned independent and dependent variables I empirically tested two different research hypotheses:

Hypothesis 1: *"Adopting better ESG practices* ($\beta_1 > 0$) *has a positive impact on Firm Perfomance explained as Return on Assets (ROA)".*

Hypothesis 2: *Good ESG policies and practices* ($\beta_1 > 0$) *guarantee in the long term an increase of Firm Performance in terms of Tobin's Q".*

Therefore, a parsimonious set of control variables provided by the academic literature has been considered including the *Size*, as the natural logarithm of Total Assets, the *Leverage*, explained by the natural logarithm of the ratio between Total debt and Enterprise Value, and the ratio between Capex and Total Assets. The regression model was empirically tested with and without the One-year lagged profitability since these values are commonly highly correlated with past values. Year and sector dummies were also considered to drop many potential bias effects. With the exception of the lagged dependent variable, all predictor variables are log-transformed by means of the natural logarithm.

To test the research hypotheses we employed the following regression models:

$$\text{1)}\quad \begin{aligned} \ln ROA_{it} = {} & \alpha + \beta_1 \ln ESG_{it} + \beta_2 \ln TotalAssets_{it} \\ & + \beta_3 \ln {}^{TotalDebt}\!/_{EV}{}_{it} + \beta_4 \ln {}^{Capex}\!/_{TotalAssets}{}_{it} \\ & + \beta_5 ROA_{(t-1)it} + \beta_6 YearDummy_{it} + \beta_7 SectorDummy_{it} + \varepsilon_{it} \end{aligned}$$

$$\text{2)}\quad \begin{aligned} \ln Tobin'sQ_{it} = {} & \alpha + \beta_1 \ln ESG_{(t-1)} + \beta_2 \ln TotalAssets_{(t-1)} \\ & + \beta_3 \ln {}^{TotalDebt}\!/_{EV}{}^{(t-1)} + \beta_4 \ln {}^{Capex}\!/_{TotalAssets}{}_{(t-1)} \\ & + \beta_5 ROA_{(t-1)it} + \beta_6 YearDummy_{it} + \beta_7 SectorDummy_{it} + \varepsilon_{it} \end{aligned}$$

Empirical Findings

Table 11.1 illustrates the descriptive statistics of the independent variables including the mean, median, maximus, minimum, and standard deviation values for the macro-categories of ESG Score and the individual pillars.

Asian companies operating in Emerging Markets show a median ESG Score of about 42, reporting a higher score for the Governance Pillar which is about 50. In particular, if we consider the individual sectors it is possible to identify which of these are more aware of ESG issues.

Table 11.1 Descriptive statistics - ESG Scores

	ESG Score	*Environmental Pillar Score*	*Social Pillar Score*	*Governance Pillar Score*
Mean	41.57	36.64	39.32	49.45
Median	40.49	34.27	36.88	49.94
Maximum	92.45	97.40	97.15	96.73
Minimum	1.12	0.00	0.05	0.73
St. Deviation	22.05	27.64	26.16	23.21

This table shows the descriptive statistics for ESG Score and individual Pillars such as Environmental, Social and Governance Scores. *Source* Thomson Reuters Refinitiv; Author's elaboration

The best Asian companies in terms of ESG Score work in Energy, Information Technology, and Consumer Services sectors, with a median score, respectively, about 53, 52 and 47.

Analysing the individual pillars, while Energy and IT companies have a better score for environmental (56 and 50) and social pillars (52 and 49), instead, the governance score is more uniform across industries. However, the higher governance score is associated with the Firms operating in Communication Services (64).

More details about descriptive statistics for individual sectors can be found in the Appendix.

Table 11.2 shows the descriptive statistics of the dependent and control variables.

In order to address more reliable results, all accounting variables have been winsorized between 95 and 5%. The main reason why we decided to winsorized variables is to offer the most accurate estimates by eliminating the presence of possible outliers. If we examine the median of dependent variables, the value of the Return on Assets (ROA) is about 11% and Tobin's Q is around 13%. Considering individual sectors, the best Firms' Performance in terms of the Return on Assets (ROA) can be traced in Communication Services, Consumer Staples and Health Care (19%,16%, 14%); while, if we analyse the median Tobin's Q across sectors the best long term performance can be traced Health Care and Consumer Staples (26% and 25%).

With regard to the control variables, the median of the Total Assets is about $ 6 million fluctuating between a maximum value of $ 200 million

Table 11.2 Descriptive statistics - ROA, Tobin's Q and control variables

	Ebitda (mln $)	*ROA (%)*	*Tobin's Q (%)*	*Total Assets (mln $)*	*Capex/Total Assets (%)*	*Leverage (%)*
Mean	1,683,434.00	12.70	17.94	12,400,000.00	5.65	3.93
Median	614,409.50	10.94	12.64	5,897,716.00	4.41	0.07
Maximum	79,200,000.00	55.95	136.59	201,000,000.00	20.58	76.52
Minimum	−3,164,990.00	−0.72	3.74	415,626.00	0.01	0.00
St. Deviation	4,534,357.00	8.70	15.82	19,900,000.00	4.54	13.74

This table shows the descriptive statistics for accounting variables by industry. *Source* Thomson Reuters Datastream; Author's elaboration

and a minimum value of $ 0.415 million. The median of Earnings Before Interest, Tax, and Depreciation (EBITDA) is around $ 0.614 million moving by a maximum value of $ 79 million to a minimum value of $ -3 million. Then, the median values of the Leverage and the ratio between Capex and Total Assets are, respectively, equal to 0.07% and 4.41%.

More details about descriptive statistics for individual sectors can be found in the Appendix.

a) Return on Assets (ROA) and ESG Scores

Starting with the first research hypothesis, I sought to empirically estimate the relationship between Return on Assets (ROA) and ESG Score. Table 11.3 shows the results for the ESG macro-categories including the ESG Score, the ESG Combined Score, and the ESG Controversies Score. Firstly, only year dummies have been introduced. For sake of readability coefficients of control variables are not displayed.[3] According to the Hausman test, the more suitable model between Fixed Effects (FE) and Random Effects (RE) is the FE model.

Estimates of the first research hypothesis point out a positive and significant link between the ESG Score and the Return on Assets (ROA). Contrary, being involved in controversial issues affects negatively the company and lowers the ESG score. However, considering the ESG Combined score, which takes into account also controversial issues, we can confirm the positive and significant impact of non-financial factors such as ESG suggesting that the development of ESG practices increases the Profitability for Asian Emerging Markets. Therefore, for a 10% increase in ESG score, the Return on Assets (ROA) value grows to 0.005[4] with a p-value equal to 0.066. Then, for a 10% increase in ESG Combined Score, the Profitability rises to 0.0048 with a p-value equal to 0.081. If we add the one-year lagged profitability into the regression model, the sign of the coefficients is confirmed, but none is significant.

Considering Sector dummies only the Random Effects (RE) model is capable of capturing both the time effect and the sectorial effect.

[3] The control variables included in the regression model are the natural logarithm of Total Assets, and the natural logarithm of Capex/Total Assets and Total Debt/Total Assets ratios.

[4] $\beta * \ln(1.1) = 0.0517 * \ln(1.1) = 0.005$.

Table 11.3 ESG Score and ROA

	3.1	*3.2*	*3.3*	*3.4*	*3.5*	*3.6*	*3.7*	*3.8*	*3.9*	*3.10*	*3.11*	*3.12*
LogESG Score	0.0517* −0.028	0.0409 −0.0268	0.0607*** −0.0233	0.0148 −0.0182								
LogESG Controversies Score					−0.0186 −0.0244	−0.0311 −0.0232	−0.0156 −0.0239	−0.0235 −0.0239				
LogESG Combined Score									0.0481* −0.0198	0.0329 −0.0263	0.0582*** −0.0231	0.0111 −0.0182
Sector dummies	NO	NO	YES	YES	NO	NO	YES	YES	NO	NO	YES	YES
Regression Model	FE	FE	RE	RE	FE	FE	RE	RE	FE	FE	RE	RE
Control variables	YES	YES	YES	YES	YES	YES	YES	YES	YES	YES	YES	YES
Constant	2.0049	1.5214	2.5925	1.3477	1.6618	1.8736	2.7493	1.4812	2.0152	1.5518	2.5882	1.3485
Observations	4013	3973	4013	3973	4013	3973	4013	3973	4013	3973	4013	3973
Number of firms	413	413	413	413	413	413	413	413	413	413	413	413
R - Squared:												
Within	0.0647	0.1735	0.0631	0.1707	0.0639	0.1734	0.0624	0.1708	0.0646	0.1733	0.0630	0.1706
Between	0.2218	0.7592	0.3513	0.7697	0.2316	0.765	0.3453	0.7702	0.2248	0.7612	0.3511	0.7700
Overall	0.1303	0.5209	0.2374	0.5336	0.1257	0.5232	0.2337	0.5339	0.1303	0.5215	0.2372	0.5337

Standard errors in brackets; standard errors are clustered on firm; ***, ** and * denotes significance at the 1%, 5% and 10% level respectively;

This table shows the results for the first research hypothesis considering the Return on Assets (ROA) as dependent variable and ESG macro-categories as independent variables. *Source* Thomson Reuters Datastream-Refinitiv; Author's elaboration

Sector dummies are dropped in the Fixed Effects (FE) model. Therefore, although the Hausman test suggests the choice of the Fixed Effects (FE) model, it was deemed more appropriate to illustrate the Random Effects (RE) model. The sign of coefficients is confirmed, but also in this case when we consider the specification of the model with the one-year lagged profitability none is significant.

Moving to individual Pillars at Environmental, Social, and Governance level, estimates show a positive link only for the Environmental and Social Pillar score, while Governance Pillar Score is negatively correlated with the dependent variable. However, the first specification of the model with Year dummies both with the one-year lagged dependent variable and without points out that the only coefficient of the Environmental Pillar score is statistically significant.

According to the Hausman test, the Fixed Effects model is more suitable. Therefore, for a 10% increase in the Environmental Pillar score, the difference in the expected mean ROA value is 0.003 with a p-value equal to 0.060. If we add the one-year lagged profitability the sign of coefficients is confirmed, but the Environmental Pillar score is less significant (p-value = 0.100).

If we consider the model specification with both year and industry dummies, the sign of coefficients is confirmed and the estimates become highly significant up to 1%. In this case, we can validate the positive impact of social practices on Profitability. Briefly, adopting good environmental and social practices allows Asian firms to increase their profitability by 0.004. If we add the one-year lagged dependent variable, the results are partially confirmed. In fact, the Environmental Pillar Score is always significant (p-value = 0.022), while the Social Pillar Score has a positive but not significant impact.

With regard to Governance Pillar Score, the coefficient is again negative but not significant. The results are illustrated in Table 11.4. The coefficients on other firm-level variables such as Size, Leverage, and the ratio between Capex and Total Assets are consistent with the existing literature.

After these initial findings, it seemed appropriate to investigate the subcategories of the Environmental, Social, and Governance Pillar score. Interesting evidence may be highlighted. I tested the regression model for all ESG subcategories. However, for sake of readability only significant coefficients are displayed.

Table 11.4 ESG Pillars Scores and ROA

	4.1	*4.2*	*4.3*	*4.4*	*4.5*	*4.6*	*4.7*	*4.8*	*4.9*	*4.10*	*4.11*	*4.12*
LogEnv Pillar Score	0.0299* (0.0159)	0.0237 (0.0152)	0.0461*** (0.0141)	0.0273** (0.0119)								
LogSocial Pillar Score					0.0245 (0.0173)	0.0167 (0.0164)	0.0427*** (0.0149)	0.0141 (0.0121)				
LogGov Pillar Score									−0.006 (0.0237)	−0.009 (0.0225)	−0.008 (0.021)	−0.0160 (0.0172)
Sector dummies	NO	NO	YES	YES	NO	NO	YES	YES	NO	NO	YES	YES
Regression Model	FE	FE	RE	RE	FE	FE	RE	RE	FE	FE	RE	RE
Control variables	YES	YES	YES	YES	YES	YES	YES	YES	YES	YES	YES	YES
Constant	1.6618	1.1683	2.4863	1.2324	2.1078	1.6223	2.6614	1.3672	2.1693	1.7517	2.6893	1.3897
Observations	3532	3498	3532	3498	4013	3973	4013	3973	4013	3973	4013	3973
Number of firms	399	399	399	399	413	413	413	413	413	413	413	413
R - Squared:												
Within	0.0556	0.165	0.0534	0.1626	0.0643	0.1732	0.0626	0.1707	0.0638	0.1729	0.0623	0.1703
Between	0.1802	0.7419	0.3509	0.7524	0.2429	0.7639	0.3569	0.7697	0.2364	0.7658	0.3455	0.7721
Overall	0.1189	0.5225	0.246	0.3555	0.1346	0.5229	0.2406	0.5337	0.1257	0.523	0.2338	0.5346

Standard errors in brackets; standard errors are clustered on firm; ***, ** and * denotes significance at the 1%, 5% and 10% level respectively;

This table shows the results for the first research hypothesis considering the Return on Assets (ROA) as the dependent variable and individual Pillars of ESG Score as independent variables. *Source* Thomson Reuters Datastream-Refinitiv; Author's elaboration

Estimates for the first specification of the model with year dummy variables show a positive and significant impact for the Resource Use Score and Workforce Score subcategories, and a significant and negative effect of the Shareholders Score. In line with what the Refinitiv methodology indicates for the ESG subcategories, the relationships can be interpreted as follows:

- The Resource Use Score reflects companies' ability to reduce the use of materials, energy or water, and to develop more eco-efficient solutions. The positive effect on Asian Firms' Profitability can be interpreted in terms of lower costs related to the adoption of circular approach and by improving supply chain management. Specifically, better use of resources increases the Profitability of 0.024.
- The Workforce Score highlights a company's commitment to its employees in ensuring safe conditions and opportunities in the work environment. In emerging markets this category seems to assume more relevance than in developed markets. Guaranteeing better job conditions and opportunities can represent an important advantage to create long-term value particularly for emerging economies. Our findings highlight that the development of policies "worker-oriented" entails not only a major healthy and safe workplace, job satisfaction, and equal opportunities including diversity, age, and culture, but also an economic advantage for Asian companies operating in emerging markets. In fact, for a 10% increase in Workforce Pillar Score, the difference in the expected mean Return on Assets (ROA) is 0.003 with a p-value equal to 0.044.
- The Shareholder Score measures the effectiveness of policies adopted in favour of shareholders, including equal treatment and the use of anti-takeover devices. However, the increase in these practices appears to reduce the profitability of Asian companies. The results show that for a 10% increase in Shareholder Score, the difference in expected average Return on Assets (ROA) is −0.003 with a p-value of 0.031. The reason for these negative effects can be traced to agency issues. Agency theory points to the equal treatment of shareholders as one of the most significant agency problems. In Emerging Market the incidence of controlling shareholders is much higher than in developed markets. Indeed, Majority shareholders may seek to obstruct policies aimed to ensure alignment between their interests

and those of minority shareholders. This problem could have an indirect negative effect on efficiency and overall business performance reflected in lower profitability.

If we consider the one-year lagged dependent variable, the Resource Use Score effect is positive but not significant (p-value = 0.150), while the Workforce Score and Shareholders Score coefficients are still positive and significant. In brief, the positive quantitative impact on Return on Assets (ROA) of a 10% increase in the Workforce Score is 0.003 (p-value = 0.056). Instead, for a 10% increase in Shareholder Score, the difference in expected average Return on Assets (ROA) is −0.002 with a p-value of 0.039.

Moving to model specification with both year and industry dummies, the sign of coefficients is confirmed and the estimates become highly significant up to 1%. Actually, this specification allows to highlight the presence of other significant subcategories such as Emissions Score, Community Score, Product Responsibility Score, and CSR Strategy Score. All of these subcategories seem to positively influence Firms' Profitability. Overall, for a 10% increase in these subcategories, the difference in expected average Return on Assets (ROA) is 0.003. Introducing practices to reduce the level of emissions, operating with respect for public health and business ethics, ensuring a better quality of services and products offered, and greater transparency in terms of disclosure increase the profitability of businesses.

If we add the one-year lagged dependent variable, the results are partially confirmed. In fact, only the coefficients of the Resource Use Score, Workforce Score, Shareholders Score, and CSR Strategy Score are still significant. Therefore, the capacity of a company to develop new environmental technologies and processes or Eco-designed products increase the Return on Assets (ROA) of 0.003. We can explain this positive impact as an indirect effect due to reduced costs and greater efficiency in production processes.

The coefficients on other firm-level variables such as Size, Leverage, and the ratio between Capex and Total Assets are consistent with the existing literature.

In the Tables 11.5 and 11.6 the coefficients of ESG subcategories are reported.

Table 11.5 ESG subcategories and ROA

	5.1	*5.2*	*5.3*	*5.4*	*5.5*	*5.6*	*5.7*	*5.8*	*5.9*	*5.10*	*5.11*	*5.12*
LogResource Use Score	0.0253* (0.0168)	0.0229* (0.0159)	0.0449*** (0.0151)	0.0283** (0.0128)								
LogWorkforce Score					0.0293** (0.0145)	0.0264** (0.0138)	0.0377*** (0.0131)	0.0162* (0.0109)				
LogShareholders Score									−0.0269** (0.0125)	−0.0245** (0.0119)	−0.0288*** (0.0118)	−0.0270*** (0.0104)
Sector dummies	NO	NO	YES	YES	NO	NO	YES	YES	NO	NO	YES	YES
Regression Model	FE	FE	RE	RE	FE	FE	RE	RE	FE	FE	RE	RE
Control variables	YES	YES	YES	YES	YES	YES	YES	YES	YES	YES	YES	YES
Constant	1.5323	0.9278	2.4076	1.047	2.0881	1.5938	2.6534	1.364	2.2312	1.8248	2.7559	1.4393
Observations	3208	3178	3208	3178	4013	3973	4013	3973	4013	3973	4013	3973
Number of firms	386	386	386	386	413	413	413	413	413	413	413	413
R -Squared:												
Within	0.052	0.1614	0.0494	0.1583	0.0649	0.1738	0.0633	0.1711	0.065	0.1739	0.0635	0.1714
Between	0.1692	0.7315	0.3545	0.757	0.2319	0.7601	0.3529	0.7691	0.2298	0.7676	0.3477	0.7725
Overall	0.1143	0.1577	0.2463	0.537	0.1312	0.5212	0.2383	0.5335	0.1248	0.524	0.2355	0.5352

Standard errors in brackets; standard errors are clustered on firm; ***, ** and * denotes significance at the 1%, 5% and 10% level respectively;

This table shows the results for the first research hypothesis considering the Return on Assets (ROA) as dependent and the subcategories of ESG Score as independent variables. *Source* Thomson Reuters Datastream-Refinitiv; Author's elaboration

Table 11.6 ESG subcategories and ROA

	6.1	*6.2*	*6.3*	*6.4*	*6.5*	*6.6*	*6.7*	*6.8*	*6.9*	*6.10*
LogEmissions Score	0.0323** (0.0146)	0.0153 (0.0154)								
LogEnv.Innovation Score			0.0316 (0.0224)	0.0323* (0.0200)						
LogCommunity Score					0.0269** (0.0120)	0.0078 (0.0129)				
LogProduct Responsibility Score							0.0275* (0.0155)	0.009 (0.0160)		
LogCSR Strategy Score									0.0365** (0.0160)	0.0219* (0.0137)
Sector dummies	YES	YES	YES	YES	YES	YES	YES	YES	YES	YES
Regression Model	FE	FE	RE	RE	FE	FE	RE	RE	RE	RE
Controlo variables	YES	YES	YES	YES	YES	YES	YES	YES	YES	YES
Constant	2.3749	1.7609	1.9188	1.3792	2.5791	1.348	2.3214	1.2308	2.8138	1.2937
Observations	3207	3175	2121	2107	3491	3901	3163	3134	2998	2972
Number of firms	383	383	302	302	413	413	385	385	380	380
R - Squared:										
Within	0.0564	0.155	0.0645	0.1344	0.0615	0.1638	0.0629	0.1576	0.0535	0.1656
Between	0.3651	0.7961	0.2707	0.6271	0.3486	0.766	0.357	0.7518	0.3656	0.7417
Overall	0.2578	0.5474	0.2588	0.5588	0.2422	0.5362	0.2403	0.5337	0.2645	0.5557

Standard errors in brackets; standard errors are clustered on firm; ***, ** and * denotes significance at the 1%, 5% and 10% level respectively;

This table shows the results for the first research hypothesis considering the Return on Assets (ROA) as dependent and the subcategories of ESG Score as independent variables. *Source* Thomson Reuters Datastream-Refinitiv; Author's elaboration

b) Tobin's Q and ESG Scores

In order to answer the second research hypothesis OLS regression using Tobin's Q as the dependent variable was estimated. Tobin's Q is commonly used as a proxy for firm value or for firm's perspectives of profitability as it is intended to capture the value of long-term investments including intangibles (Bellavite Pellegrini et al., 2020). In line with the existing literature, all predictor variables are one-year lagged.

The following estimates strengthen the effect of the coefficients observed above. In general, we can observe a positive and significant impact of ESG Score and ESG Combined Score even when we add one-year lagged dependent variable. In this case, the coefficient of the ESG Controversies Score also appears positive but not significant.

Table 11.7 reports the results for the one-year lagged ESG macro-categories. Firstly, only year dummies are implemented. According to the Hausman test, the Fixed Effects (FE) model is more suitable than the Random Effects model (RE). Therefore, for a 10% increase the one-year lagged ESG Score and the one-year lagged ESG Combined Score, the difference in Tobin's Q is equal for both to 0.001, with a p-value highly significant. If we add in the regression model one-year lagged profitability the sign of the coefficients is confirmed. Turning to the second specification of the model which takes into account also Sector dummies, the results are confirmed.

Regarding the ESG Pillar Scores, we confirm the strong positive link between the Environmental and Social pillar scores and the dependent variable. However, considering the one-year lagged dependent variable, we can highlight that the Governance Score also has a positive and statistically significant effect on firm profitability (p-value = 0.100).

In Emerging Markets, governance issues still appear to be a key challenge. Therefore, the benefit of developing such practices may only be verifiable in the medium-long term. Considering the Sector dummies, the results are confirmed for the environmental and social pillar score, while the governance score coefficient is still positive but not significant.

The coefficients on other firm-level variables such as Size, Leverage and the ratio between Capex and Total Assets are consistent with the existing literature.

The results for ESG individual Pillars Scores are illustrated in Table 11.7.

Table 11.7 ESG Score and Tobin's Q

	7.1	*7.2*	*7.3*	*7.4*	*7.5*	*7.6*	*7.7*	*7.8*	*7.9*	*7.10*	*7.11*	*7.12*
LogESG Score(t-1)	**0.0149***** (0.0051)	**0.0125***** (0.0049)	**0.0178***** (0.0050)	**0.0153***** (0.0048)								
LogESG Controv. Score(t-1)					0.0023 (0.0037)	0.0050 (0.0036)	0.0035 (0.0037)	0.0019 (0.0036)				
LogESG Combined Score(t-1)									**0.0143***** (0.0051)	**0.0119***** (0.0048)	**0.0173***** (0.005)	0.0148 (0.004)
Sector dummies	NO	NO	YES	YES	NO	NO	YES	YES	NO	NO	YES	YES
Regression Model	FE	FE	RE	RE	FE	FE	RE	RE	FE	FE	RE	RE
Control variables	YES	YES	YES	YES	YES	YES	YES	YES	YES	YES	YES	YES
Constant	4.2038	4.0912	4.4792	4.1338	4.2585	4.1418	4.4715	4.111	4.2074	4.0945	4.3925	4.1326
Observations	4524	4521	4524	4521	4524	4521	4524	4521	4524	4521	4524	4521
Number of firms	414	414	414	414	414	414	414	414	414	414	414	414
R - Squared:												
Within	0.237	0.3076	0.2365	0.3051	0.2355	0.3066	0.2351	0.3039	0.2369	0.3076	0.2364	0.305
Between	0.2502	0.5576	0.4187	0.6267	0.2333	0.5484	0.413	0.6241	0.2493	0.5576	0.4185	0.6266
Overall	0.2258	0.4724	0.3794	0.5527	0.2171	0.467	0.3747	0.5506	0.2254	0.4724	0.3793	0.5526

Standard errors in brackets; standard errors are clustered on firm; ***, ** and * denotes significance at the 1%, 5% and 10% level respectively;

This table shows the results for the second research hypothesis considering the Tobin's Q as dependent variable and ESG macro-categories as independent variables.

Source Thomson Reuters Datastream-Refinitiv; Author's elaboration

Table 11.8 ESG Pillars Scores and Tobin's Q

	8.1	*8.2*	*8.3*	*8.4*	*8.5*	*8.6*	*8.7*	*8.8*	*8.9*	*8.10*	*8.11*	*8.12*
LogEnv Pillar Score(t-1)	**0.0122***** (0.0048)	**0.0102**** (0.0045)	**0.0152***** (0.0047)	**0.0124***** (0.0047)								
LogSocial Pillar Score(t-1)					**0.0165***** (0.0047)	**0.0142***** (0.0045)	**0.0196***** (0.0046)	**0.0166***** (0.0044)				
LogGov Pillar Score (t-1)									0.0054 (0.0046)	0.0042 (0.0044)	**0.0075*** (0.0046)	**0.0068*** (0.0044)
Sector dummies	NO	NO	YES	YES	NO	NO	YES	YES	NO	NO	YES	YES
Regression Model	FE	FE	RE	RE	FE	FE	RE	RE	FE	FE	RE	RE
Control variables	YES	YES	YES	YES	YES	YES	YES	YES	YES	YES	YES	YES
Constant	4.0989	3.9499	4.3925	4.0216	4.2234	4.1072	4.5004	4.1522	4.2439	4.1249	4.4722	4.1182
Observations	4030	4035	4038	4035	4524	4521	4524	4521	4524	4521	4524	4521
Number of firms	414	414	414	414	414	414	414	414	414	414	414	414
R - Squared:												
Within	0.2415	0.3157	0.2408	0.313	0.2377	0.3082	0.2372	0.3057	0.2357	0.3067	0.2352	0.3041
Between	0.274	0.5679	0.4346	0.6286	0.255	0.5578	0.421	0.6264	0.2374	0.5516	0.4141	0.6251
Overall	0.2246	0.4803	0.3761	0.5583	0.2295	0.4734	0.3814	0.5528	0.2192	0.4689	0.3757	0.5513

Standard errors in brackets; standard errors are clustered on firm; ***, ** and * denotes significance at the 1%, 5% and 10% level respectively;

This table shows the results for the second research hypothesis considering the Tobin's Q as dependent variable and individual Pillars of ESG Score as independent variables.

Source Thomson Reuters Datastream-Refinitiv; Author's elaboration

Analysing the subcategories of the Environmental, Social, and Governance Pillar Score, it is possible to observe a positive and significant impact of several categories including Resource Use Score, Emission Score, Workforce Score, Community Score, Productivity Score, and Shareholders Score. The estimates reported in Tables 11.9 and 11.10 show that when the dependent variable is captured by the natural logarithm of Tobin's Q all variables have a positive effect.

These results are confirmed both in the first specification of the model with only the year dummies and when considering the year and sector dummies.

Summarizing the results we can observe that when Asian companies operating in Emerging Markets adopt sustainability practices at the Environmental level, as well as a better use resources and limit the level of CO2 emissions, the one-year lagged independent variables increase Firms' Profitability of 0,002. We consider that the number of positive evidences related to Environmental Pillar is strongly related to the huge growth in environmental regulation. In the last ten years, different countries among Asian Emerging Markets, as well as China, have given increasing attention to the environment. This process, started in the late 1990s, after unprecedented growth in industrial manufacturing, pushed the Chinese central government to strengthen regulation on environmental issues (He et al., 2020).

Another key tool for emerging economies is certainly represented by the development of policies at the social level. In fact, almost every subcategory of the Social Pillar Score tends to have a positive impact on long-term profitability. In brief, implementing best practices aimed at ensuring greater benefits for the work environment and surrounding community, and providing better quality products increases Tobin's Q by 0.003.

The different signs of Governance Pillar Score can be traced to the positive impact of Shareholders Score subcategory. Indeed, when the dependent variable is measured by Return on Assets (ROA), the development of policies in favour of Shareholders seems to reduce the profitability, while using Tobin's Q as the dependent variable the effect is opposite. This finding highlights that an initial cost can translate into a long-term benefit. Greater alignment of interests across different types of shareholders, more efficient mechanisms for protecting minority shareholders, and more responsibility entrusted to the board lead to greater

Table 11.9 ESG(t-1) subcategories and Tobin's (Part I)

	9.1	*9.2*	*9.3*	*9.4*	*9.5*	*9.6*	*9.7*	*9.8*	*9.9*	*9.10*	*9.11*	*9.12*
LogResource Use Score (t-1)	**0.0108**** (0.0048)	**0.0093**** (0.0046)	**0.0148***** (0.0048)	**0.0127***** (0.0046)								
LogEmissions Score(t-1)					**0.0111**** (0.0048)	**0.0098**** (0.0045)	**0.0137***** (0.0047)	**0.0114***** (0.0046)				
LogWorkforce Score(t-1)									**0.0115***** (0.0042)	**0.0095***** (0.0039)	**0.0139***** (0.0041)	**0.0116***** (0.0039)
Sector dummies	NO	NO	YES	YES	NO	NO	YES	YES	NO	NO	YES	YES
Regression Model	FE	FE	RE	RE	FE	FE	RE	RE	FE	FE	RE	RE
Control variables	YES	YES	YES	YES	YES	YES	YES	YES	YES	YES	YES	YES
Constant	4.1267	4.0378	4.424	4.0722	4.1864	4.0335	4.4899	4.0892	4.2242	4.1089	4.4851	4.1373
Observations	4524	3713	3716	3713	3715	3712	3715	3712	4524	4521	4524	4521
Number of firms	414	413	413	413	412	412	412	412	414	414	414	414
R - Squared:												
Within	0.242	0.3099	0.2412	0.3068	0.245	0.3299	0.2442	0.3271	0.2369	0.03075	0.2364	0.305
Between	0.2758	0.546	0.4361	0.621	0.2573	0.5488	0.44	0.6317	0.2479	0.5561	0.4181	0.6258
Overall	0.2397	0.4819	0.3872	0.5642	0.2231	0.486	0.3812	0.5706	0.2249	0.4716	0.379	0.552

Standard errors in brackets; standard errors are clustered on firm; ***. ** and * denotes significance at the 1%. 5% and 10% level respectively;

This table shows the results for the second research hypothesis considering the Tobin's Q as dependent variable and subcategories of ESG Score as independent variables.

Source Thomson Reuters Datastream-Refinitiv; Author's elaboration

Table 11.10 ESG(t-1) subcategories and Tobin's Q (Part II)

	10.1	*10.2*	*10.3*	*10.4*	*10.5*	*10.6*	*10.7*	*10.8*	*10.9*	*10.10*	*10.11*	*10.12*
LogCommunity Score (t-1)	**0.013***** (0.0042)	**0.0117***** (0.0041)	**0.0158***** (0.0042)	**0.0137***** (0.0046)								
LogProducts Responsibility Score(t-1)					**0.0117***** (0.0041)	**0.0095**** (0.0045)	**0.0145***** (0.0047)	**0.0116***** (0.0045)				
LogShareholders Score(t-1)									**0.0075**** (0.0038)	**0.0068**** (0.0037)	**0.0088***** (0.0038)	**0.0087***** (0.0037)
Sector dummies	NO	NO	YES	YES	NO	NO	YES	YES	NO	NO	YES	YES
Regression Model	FE	FE	RE	RE	FE	FE	RE	RE	FE	FE	RE	RE
Control variables	YES	YES	YES	YES	YES	YES	YES	YES	YES	YES	YES	YES
Constant	4.1597	4.0399	4.4464	4.1045	4.3805	4.2371	4.6367	4.2206	4.2241	4.101	4.4775	4.1211
Observations	4449	4446	4449	4446	3682	3679	3682	3679	4524	4521	4524	4521
Number of firms	414	414	414	414	411	411	411	411	414	414	414	414
R - Squared:												
Within	0.2388	0.3094	0.2383	0.3071	0.2368	0.315	0.2359	0.3108	0.2361	0.3071	0.2357	0.3045
Between	0.2593	0.5628	0.4217	0.625	0.2373	0.518	0.4259	0.6213	0.2387	0.5529	0.4145	0.626
Overall	0.2309	0.4782	0.3842	0.5555	0.221	0.4716	0.3815	0.5691	0.2204	0.47	0.3761	0.5521

Standard errors in brackets; standard errors are clustered on firm; ***, ** and * denotes significance at the 1%, 5% and 10% level respectively;

This table shows the results for the second research hypothesis considering the Tobin's Q as dependent variable and subcategories of ESG Score as independent variables.

Source Thomson Reuters Datastream-Refinitiv; Author's elaboration

profitability over time, even in markets characterized by weaker legal systems such as emerging markets.

The coefficients on other firm-level variables such as Size, Leverage, and the ratio between Capex and Total Assets are consistent with the existing literature.

The results for one-year lagged ESG subcategories are reported in Table 11.10.

Conclusion

The aim of this research work is to study the relationship between Financial Performance and the development of Environmental, Social, and Governance practices in Asian Emerging Markets. In particular, considering a sample of 415 Asian companies operating in emerging markets between 2008 and 2019, I attempted to analyse how the adoption of environmental, social, and governance (ESG) practices could influence Firms' Profitability.

The literature on the relationship between corporate financial performance and extra-financial performance historically focused on developed markets (Bassen et al., 2006; Cheng et al., 2014; Dhaliwal et al., 2011). However, although the path towards sustainability is slower in Emerging Markets than in Developed markets, there are several studies that highlight the existence of a "*sustainability business case*" for Emerging Markets. In fact, sustainability does not seem to place limits on development, but rather seems able to create new opportunities for growth and economic development. Therefore, this research study seeks to contribute to the academic literature analysing which could be the most important action sustainability areas for Asian Emerging Markets.

In order to assess the link between ESG practices and Firms' Profitability two research hypotheses were defined. The first research hypothesis tests how adopting good ESG practices impacts the Return on Asset (ROA), while the second one takes into consideration another proxy commonly used to measure firm value or for firm's perspectives in the long-term as well as the Tobin's Q.

Starting with the first research hypothesis which considers the Return on Assets (ROA) as the dependent variable, we can confirm the positive relationship between the ESG Score and Firm Profitability up to 1%. Even when we consider the ESG Combined Score, which takes into account controversies issues, a positive and significant impact can be observed. Again, being involved in controversial issues affects negatively the company and lowers the ESG score.

The ESG Score individual pillars analysis points out a positive and significant impact of the Environmental and Social Pillars scores. The development of greater environmental practices always increases the Firm's profitability, while adopting social policies seems to affect in a statistically significant way as we consider both year and sector dummies.

Through the ESG methodology of Refinitiv we had the opportunity to analyse in detail each of the subcategories of the individual Pillars. Estimates for the first specification of the model highlight a positive and significant impact of the Resource Use Score and Workforce Score, and a significant and negative effect of the Shareholders Score.

Therefore, being more aware of the use of resources for adopting a better approach in terms of the environmental impact increases the Asian Firm's Profitability. Moreover, Asian companies of Emerging Markets tend to be rewarded for social policies implementation in favour of workers. While the negative effect of Shareholders Score subcategory can be traced to the lack of alignment between the interests of controlling shareholders and minority shareholders. This issue, which represents one of the most important agency problems in Corporate Governance, strongly characterizes Emerging Markets where the incidence of controlling shareholders is much higher than developed markets.

Moving to the model specification with the one-year lagged dependent variable, the Resource Use Score effect is positive but not significant, while the Workforce Score and Shareholders Score coefficients are still positive and significant. Considering both year and industry dummies, we highlight the presence of other significant subcategories such as Emissions Score, Community Score, Product Responsibility Score, and CSR Strategy Score. All of these seem to positively influence Firms' Profitability. Therefore, the introduction of practices to reduce the level of emissions, operating with respect for public health and business ethics, ensuring a better quality of services and products offered, and greater transparency in terms of disclosure increase the Firms' Profitability in Asian Emerging Markets.

If we add the one-year lagged dependent variable, the results are partially confirmed. But we can observe that even the Environmental Innovation Score, as well as the capacity of a company to implement new environmental technologies and processes or Eco-designed products, increases the Return on Assets (ROA).

With regard to the second research hypothesis, which considers Tobin's Q as proxy of profitability is confirmed blustering results observed previously. Tobin's Q is commonly used as proxy for firm value or for firm's perspectives of profitability as it is intended to capture the value of long-term investments including intangibles (Bellavite Pellegrini et al., 2020). In this case, in line with the existing literature, all predictor variables are one-year lagged.

The results strengthen the previous considerations. The positive and significant impact of the one-year lagged ESG Score and ESG Combined Score is confirmed in all specifications of the model, even as we add one-year lagged dependent variable. In this case, the coefficient of the ESG Controversies Score also appears positive but not significant.

Regarding the ESG Pillar Scores, we confirm the strong positive link between the Environmental and Social pillar scores and the dependent variable. However, considering the one-year lagged dependent variable, we can see that the Governance Score also has a positive and statistically significant effect on firm profitability (p-value = 0.100).

The analysis of the subcategories of the Environmental, Social, and Governance Pillar Score, highlights a positive and significant impact of several categories including Resource Use Score, Emission Score, Workforce Score, Community Score, Productivity Score and Shareholders Score. Despite previous findings, when the dependent variable is captured by the natural logarithm of Tobin's Q, all variables have a positive effect. These results are confirmed both in the first specification of the model with only the year dummies and when considering the year and sector dummies.

Estimates suggest that when Asian companies adopt sustainability practices at the Environmental level, as well as a better use resources and limit the level of CO2 emissions, the value of Tobin's Q increased. In the last ten years, different countries among Asian Emerging Markets, as well as China, have given increasing attention to environmental reinforcing several aspects of Regulation. This process, started in the late 1990s, seems to have led over time to benefits not only in favour of the environment but also in terms of performance for the most virtuous companies.

Moreover, another important aspect for emerging economies is represented by the Social Pillar. In fact, almost every subcategory of the Social Pillar Score tends to have a positive impact on long-term Firms' Profitability. In brief, implementing best practices aimed at ensuring greater benefits for the work environment and surrounding community, and providing better quality products increases Tobin's Q by 0.003.

The different signs of Governance Pillar Score can be traced to the positive impact of Shareholders Score subcategory. In Emerging Markets, governance issues still appear to be a key challenge. However, it is possible to highlight benefits in the medium-long term. Greater alignment of interests across different types of shareholders, more efficient mechanisms for protecting minority shareholders, and more responsibility entrusted to the board lead to greater profitability over time, even in markets characterized by weaker legal systems such as emerging markets.

Finally, we can conclude that the adoption of good ESG practices and policies can be an important advantage for Asian firms operating in emerging markets. In line with previous evidence in the literature, this empirical work confirms that the main areas of sustainability action for emerging economies appear to be improving environmental processes and practices and human resource management. However, we highlight that efforts to develop good governance practices, such as greater alignment between the interests of different categories of shareholders, can also reward companies in the long run. In this context, the role of government is crucial. We believe that growing awareness and importance by governments among environmental issues have prompted emerging markets to adopt more effective practices and policies. Future studies could investigate whether a causality relationship exists to support an increasingly tangible commitment to the implementation of ESG practices in Asian Emerging Markets as a *win-win* to both companies and their entire environment.

Appendix

See Tables 11.11, 11.12, 11.13

Table 11.11 ESG Score, ESG Pillars Scores -Descriptive statistics by industry

	ESG Score	*Environmetal Pillar Score*	*Social Pillar Score*	*Governance Pillar Score*
Communication Services				
Mean	46.135	30.701	42.982	60.499
Median	46.855	26.943	40.735	64.0667
Maximum	87.546	80.304	91.696	94.195
Minimum	9.527	0.00	2.121	6.4102
Standard Deviation	18.699	23.948	23.028	20.751
Consumer Discretionary				
Mean	36.284	32.695	34.147	42.374
Median	31.889	26.029	27.263	40.505
Maximum	92.365	96.891	96.831	92.089
Minimum	1.911	0.00	0.222	1.0029
Standard Deviation	23.064	29.748	26.401	23.111
Consumer Staples				
Mean	42.521	36.898	42.099	48.745
Median	39.901	32.735	38.775	49.845
Maximum	86.384	94.577	96.487	88.056
Minimum	2.957	0.00	1.786	4.633
Standard Deviation	22.532	28.879	26.405	21.604
Energy				
Mean	51.601	51.172	51.061	51.910
Median	53.119	55.897	52.550	50.842
Maximum	88.276	94.133	97.153	94.982
Minimum	5.995	0.00	1.399	6.048
Standard Deviation	19.265	24.341	24.255	22.600
Health Care				
Mean	35.825	18.352	33.072	51.723
Median	28.689	10.131	29.170	44.904
Maximum	85.858	69.149	91.852	96.734
Minimum	4.683	0.00	1.341	12.500
Standard Deviation	20.906	20.131	24.487	22.878
Industrials				
Mean	36.521	33.439	32.682	45.419
Median	34.639	30.479	28.295	46.309

(continued)

Table 11.11 (continued)

	ESG Score	*Environmetal Pillar Score*	*Social Pillar Score*	*Governance Pillar Score*
Maximum	83.025	96.549	93.801	92.141
Minimum	1.239	0.00	0.0529	1.259
Standard Deviation	19.876	26.137	23.903	22.137
Information Technology				
Mean	50.718	47.345	47.859	58.488
Median	52.723	49.953	49.407	60.836
Maximum	92.454	97.401	95.675	96.120
Minimum	3.697	0.00	0.229	2.868
Standard Deviation	24.401	28.224	28.462	23.896
Materials				
Mean	39.568	38.508	35.781	47.625
Median	39.409	37.381	30.672	49.036
Maximum	85.814	94.595	95.948	95.482
Minimum	1.119	0.00	0.284	0.728
Standard Deviation	22.687	27.564	27.372	22.658
Real Estate				
Mean	35.575	24.592	37.651	44.236
Median	33.107	12.590	33.945	45.019
Maximum	85.309	86.966	92.435	89.453
Minimum	2.508	0.00	0.890	1.542
Standard Deviation	20.044	2.189	22.827	23.008
Utilities				
Mean	42.577	37.553	43.288	48.934
Median	42.402	36.189	39.071	49.663
Maximum	82.033	92.498	88.658	94.674
Minimum	6.893	0.00	8.203	5.813
Standard Deviation	16.456	20.419	22.959	21.291

This table shows the descriptive statistics for ESG Score and individual Pillars such as Environmental, Social and Governance Scores by industry.
Source Thomson Reuters Datastream-Refinitiv; Author's elaboration.

Table 11.12 Accounting variables—Descriptive statistics by industry

	Ebitda (millions of $)	*ROA (%)*	*Tobin's Q (%)*	*Total Assets (millions of $)*	*Capex/Total Assets (%)*	*Leverage (%)*
Communication Services						
Mean	3,472,599.00	21.491	24.217	15,300,000.00	9.294	0.076
Median	1,313,417.00	19.087	17.336	8,017,518.00	8.712	0.058
Maximum	45,700,000.00	48.579	76.578	89,200,000.00	18.490	0.236
Minimum	−74,223.00	7.738	7.803	597,671.00	1.506	0.0002
Standard Deviation	7,159,625.00	9.795	18.641	21,100,000.00	4.822	0.0658
Consumer Discretionary						
Mean	996,068.50	13.327	19.409	8,293,492.00	5.201	3.339
Median	437,990.50	11.927	14.734	3,627,599.00	4.264	0.069
Maximum	13,800,000.00	31.018	53.770	36,000,000.00	14.309	37.599
Minimum	−607,594.00	2.485	4793	546,619.00	0.835	0.0001
Standard Deviation	1,628,898.00	7.536	13.527	10,200,000.00	3.637	9.774
Consumer Staples						
Mean	606,134.80	19.656	38.833	3,969,305.00	6.445	0.063
Median	447,496.50	16.431	25.183	3,022,103.00	5.299	0.043
Maximum	3,252,797.00	55.954	136.599	11,700,000.00	15.731	0.238
Minimum	−490,143.00	1.956	5.056	458,312.00	1.220	0.0001
Standard Deviation	525,700.70	13.471	35.464	3,204,759.00	3.925	0.069
Energy						
Mean	4,984,267.00	13.822	14.718	30,900,000.00	6.979	0.075
Median	1,275,986.00	12.717	12.533	11,300,000.00	5.954	0.055
Maximum	57,700,000.00	31.233	30.969	201,000,000.00	16.233	0.276
Minimum	−1,590,025.00	1.310	73.210	1,143,282.00	1.351	0.0002
Standard Deviation	9,519,822.00	7.638	6.410	48,600,000.00	4.285	0.070
Health Care						
Mean	419,453.80	15.549	30.401	3,274,289.00	4.968	0.159
Median	265,495.50	14.605	26.512	1,841,841.00	4.988	0.027
Maximum	2,783,373.00	27.886	70.236	19,000,000.00	10.932	2.047
Minimum	−17,385.00	6.502	3.743	504,358.50	0.721	0.00
Standard Deviation	460,845.40	6.308	18.456	4,389,962.00	2.943	0.457
Industrials						
Mean	938,383.20	7.755	12.538	12,400,000.00	36.471	10.244

(continued)

Table 11.12 (continued)

	Ebitda (millions of $)	*ROA (%)*	*Tobin's Q (%)*	*Total Assets (millions of $)*	*Capex/Total Assets (%)*	*Leverage (%)*
Median	511,042.00	7.362	11.030	7,511,482.00	25.265	0.11
Maximum	13,200,000.00	17.036	25.642	53,700,000.00	1.153	76.521
Minimum	−3,164,990.00	−0.717	6.200	1,000,304.00	0.396	0.0017
Standard Deviation	1,537,072.00	4.488	4.839	13,700,000.00	3.092	2.290
Information Technology						
Mean	1,915,154.00	13.503	15.798	7,264,936.00	6.544	1.706
Median	446,467.50	11.989	12.199	4,347,113.00	4.552	0.034
Maximum	79,200,000.00	3.264	43.452	29,300,000.00	20.579	21.426
Minimum	−639,632.00	0.628	5.446	415,626.00	0.339	0.00
Standard Deviation	6,578,147.00	8.390	9.836	7,803,210.00	5.901	5.485
Materials						
Mean	1,318,669.00	12.096	15.361	10,700,000.00	6.026	7.152
Median	676,892.50	10.843	12.430	6,424,348.00	4.726	0.076
Maximum	16,500,000.00	27.392	42.798	39,900,000.00	15.204	60.791
Minimum	−2,169,102.00	0.232	6.245	813,578.00	1.185	0.0004
Standard Deviation	2,082,795.00	6.923	8.974	10,700,000.00	3.919	17.346
Real Estate						
Mean	1,298,749.00	7.281	11.659	15,300,000.00	1.867	0.134
Median	497,932.00	6.652	10.598	7,079,914.00	0.621	0.092
Maximum	20,100,000.00	15.856	22.907	82,700,000.00	10.329	0.594
Minimum	−741,602.00	2.146	5.507	1,507,877.00	0.009	0.019
Standard Deviation	2,499,604.00	3.636	4.549	20,600,000.00	2.866	0.134
Utilities						
Mean	2,051,380.00	10.305	13.572	22,700,000.00	7.569	0.154
Median	1,074,154.00	9.868	11.856	11,800,000.00	6.428	0.120
Maximum	24,700,000.00	19.127	25.968	95,400,000.00	18.908	0.371
Minimum	−231,13.00	3.488	7.721	2,335,867.00	0.936	0.019
Standard Deviation	2,677,391.00	4.273	5.0168	23,700,000.00	5.003	0.104

This table shows the descriptive statistics for accounting variables by industry.
Source Thomson Reuters Datastream-Refinitiv; Author's elaboration.

Table 11.13 Pearson Correlation

	ROA	*ESG Score*	*Total Assets*	*Capex/Total Assets*	*Leverage*
ROA	1				
ESG Score	0.1422*	1			
Total Assets	−0.1098*	0.1691*	1		
Capex/Total Assets	0.3287*	0.1296*	0.0763*	1	
Leverage	−0.1813*	0.0575*	0.0559*	−0.0688*	1
*denotes significance at 5% level					

This table shows the Pearson Correlation implemented through STATA IC 13; Authors' elaboration. *Source* Thomson Reuters Datastream-Refinitiv; Author's elaboration.

References

Bassen, A., Meyer, K., & VSchlange, J. (2006). The influence of corporate responsibility on the cost of capital. Working paper series. University of Hamburg.

Baughn, C. C., Bodie, N. L. D., & Mcintosh, J. C. (2007). Corporate social and environmental responsibility in Asian countries and other geographical regions. *Corporate Social Responsibility and Environmental Management, 14*(4), 189–205.

Bellavite Pellegrini, C., Caruso, R., & Cifone, R. (2020). The impact of ESG scores on both firm profitability and value in the tyre and automotive sector (Working paper).

Cheng, B., Ioannou, I., & Serafeim, G. (2014). Corporate social responsibility and access to finance. *Strategic Management Journal, 35*, 1–23.

Dhaliwal, D. S., Li, O. Z., Tsang, A., & Yang, Y. G. (2011). Voluntary nonfinancial disclosure and the cost of equity capital: The initiation of corporate social responsibility Reporting. American Accounting Association. *The Accounting Review, 86*, 59–100.

Dobers, P., & Halme, M. (2009). Corporate social responsibility and developing countries. *Corporate Social Responsibility and Environmental Management, 16*(5), 237–249.

Elkington, J. (1997). *Cannibals with forks: The triple bottom line of 21st century business*. Capstone.

Farooq, O. (2013). Financial centers and the relationship between ESG disclosure and firm performance: Evidence from an emerging market. *The Journal of Applied Business Research, 31*(4), 1239–1244.

Friede, G., Busch, T., Bassen, A. (2015). ESG and financial performance: Aggregated evidence from more than 2000 empirical studies. *Journal of Sustainable Finance & Investment*, pp. 210–233.

Garcia, A. S., Medes-Da-Silva, W., & Orsato, R. J. (2017). Sensitive industries produce better ESG performance: Evidence from emerging markets. *Journal of Cleaner Production*, pp. 135–147.

He, G., Wang, S., & Zhang, B. (2020). Watering down environmental regulation in China. *The Quarterly Journal of Economics, 135*(4), 2135–2185.

Jensen, M. C., & Meckling, W. H. (1976). Theory of the firm: Managerial behavior, agency costs and ownership structure. *Journal of Financial Economics, 3*(4), 305–360.

Lang, M. H., Lins, K. V., & Miller, D. P. (2004). Concentrated control, analyst following, and valuation: Do analysts matter most when investors are protected least? *Journal of Accounting Research, 42*(3), 589–623.

Odell, J., & Ali, U. (2016). ESG Investing in emerging and frontier markets. *The Journal of Applied Corporate Finance, 28*(2), 96–101.

Thorpe, J., Prakash-Mani, K. (2003). Developing value: The business Case for Sustainability in Emerging Markets. Sustainability, UK.

UN report. (2019). The sustainable development goals report 2019. United Nations. https://unstats.un.org/sdgs/report/2019/The-Sustainable-Development-Goals-Report-2019.pdf

Index

© The Editor(s) (if applicable) and The Author(s), under exclusive license to Springer Nature Switzerland AG 2022

C. B. Pellegrini et al. (eds.), *Climate Change Adaptation, Governance and New Issues of Value*, Palgrave Studies in Impact Finance,
https://doi.org/10.1007/978-3-030-90115-8

The manufacturer's authorised representative in the EU is Springer Nature Customer Service Centre GmbH, Europaplatz 3, 69115 Heidelberg, Germany. If you have any concerns regarding our products, please contact ProductSafety@springernature.com

Printed and bound by CPI Group (UK) Ltd, Croydon, CR0 4YY
30/09/2026
02225321-0011